Ethics in Cyberspace

Thomas Ploug

Ethics in Cyberspace

How Cyberspace May Influence
Interpersonal Interaction

 Springer

Professor Thomas Ploug
Copenhagen Institute of Technology, AAUK.
Lautrupvang 2B
2750 Ballerup
Denmark
ploug@hum.aau.dk
post@thomasploug.dk
http://thomasploug.dk

Artwork used in the cover design courtesy of Alex Gomez and Kristian Nørgaard

ISBN 978-90-481-8492-7   e-ISBN 978-90-481-2370-4
DOI 10.1007/978-90-481-2370-4
Springer Dordrecht Heidelberg London New York

Printed on acid-free paper

Springer is part of Springer Science+Business Media (www.springer.com)

Acknowledgements

In the process of writing this book and the doctoral dissertation on which it is based I have enjoyed various forms of assistance from a number of people. For all of this I am very grateful. Among friends and colleagues I would like to thank: Peter Øhrstrøm, for his encouragement, interesting conversation and valuable comment; Klaus Robering, for his valuable comments on extensive parts of the book; Kasper Lippert-Rasmussen, for very clear, competent and balanced criticism on the first parts of the book; and Søren Holm, for the hospitality shown during my stay as visiting research fellow at the University of Cardiff, and for his clear, balanced and insightful comment on the final product. I also owe thanks to Springer for seeing the potential in the subject of this book, and in particular to Natalie Rieborn and Neil Olivier. Thanks, too, to the two anonymous reviewers at Springer for their thorough comments and suggestions. Finally I owe my deepest gratitude to my family: to Mum and Dad, for your tireless interest in my work; to Berit, for your love and patience throughout: to Naomi and Celine, for the smiles and laughter that can turn any day to sunshine.

Contents

Part I

The basic premise

Chapter 1

Ethics in cyberspace

Over recent decades information technology has come to have an increasingly prominent role in our dealings with other people. The computer, in particular, has made
available a host of new ways of interacting, which we have increasingly been able to
use to our advantage. In the wake of this development a number of ethical questions
have been raised and debated.

This book focuses on the possible consequences for moral agency of mediating
interaction by means of computers. It seeks to clarify how the conditions governing
certain kinds of interaction in cyberspace differ from face-to-face interaction, and how
this difference may come to affect the behaviour of interacting agents in a way that
has relevance for ethics. More specifically, the book endeavours to shed light on some
of the factors influencing our conviction that a particular other person is real, to
suggest how this conviction may be affected by moving the setting of interaction from
outside to inside cyberspace, and finally to show how these changes may lead an agent
to behave differently, ethically speaking, in the two settings.

In the process of addressing the topics mentioned we will focus our attention more
specifically upon interaction in ordinary text-based chat-rooms, virtual worlds such as
Second-Life and also, to a certain degree, telerobotics. As will become clear, however,
the inquiries have a broader application. They may also tell at least part of the story of
how, more generally, spatial and temporal distance may come to affect the behaviour
of interacting agents in an ethically relevant way.

T. Ploug, *Ethics in Cyberspace: How Cyberspace May Influence Interpersonal Interaction,*   **3**
© Springer Science+Business Media B.V. 2009

1.1   Introduction

1.1.1   The face of the other

The genesis of this book lies in part in a re-reading of some of the works of the French philosopher Emmanuel Levinas. Prompted by an interest in how the mediation of interaction by means of computers may come to influence agency in an ethically relevant way,[1] it seemed an obvious step to begin the investigation by exploring the works of Levinas. A very brief summary of his ideas may clarify this.

At the core of Levinas' ethical thinking lies the notion of 'the face of the other'. It is through the transcendent face of the other that we enter into a relationship with the infinite other.[2] The transcendent face of the other has meaning content that is independent of context and expresses the appeal, the requirement and the commandment expressed by: 'Thou shall not kill'.[3] According to Levinas, this appeal, requirement or commandment constitutes a fundamental responsibility in and through which the self comes into being.[4] Without going further into Levinas' thinking, it seems as though it may be interpreted as implying the existence of an ethically relevant difference between interaction in cyberspace and interaction face-to-face. Thus, given that the requirement of the transcendent face of the other constitutes not only human existence itself but also our specific moral responsibility for the other,[5] and given that the transcendent face is somehow tied to the presence of the bodily face of the other,[6] then, other things being equal, there is reason to suppose that ultimately the behaviour of interacting agents may be affected in an ethically relevant way simply because that interaction takes place in cyberspace. After all, certain kinds of interaction in cyberspace involve the loss of perceptual access to the bodily face of the interacting party.

---

[1] Note that the concepts of morality and ethics are used interchangeably.

[2] Levinas in [55, pp. 100 and 103].

[3] Levinas in [55, pp. 100–101, 103 and 114], and in [56, p. 195].

[4] Levinas in [55, pp. 111–112], and in [57, p. 206].

[5] Although Levinas is primarily concerned with the basic conditions of *human existence*, he takes his considerations to have normative ethical implications. Cf. [55, p. 105].

[6] Central to Levinas' ethical thinking is the idea that another person is infinite, irreducible 'otherness'. The other escapes any attempt of conceptual representation available to the mind. Hence the notion of a 'transcendent face'. To 'see' the ethical content of the transcendent face thus requires 'a meeting' – and this seems to presuppose the presence of the bodily face. Moreover, Levinas seems to link the ethical content of the transcendent face to the vulnerability and poverty of the bodily face. Cf. [57, pp. 44–45 and 110–112].

Intuitively Levinas' thinking seems to capture something important and uncontroversial. Being face-to-face with another person does appear to influence the character of those of our actions which affect that other person. Whether or not this is to be explained by invoking notions such as a transcendent face is, however, more controversial. Here our approach diverges from that of Levinas. While Levinas' observation that the face of the other is of importance for ethics forms the starting-point of this book, in what follows we will develop an alternative account of his observation. Accordingly, this book may be seen as an investigation into the role of the bodily face for the ethical character of interaction, where 'bodily face' is to be taken in a rather broad sense. This book has implications, therefore, that go beyond cyberspatial interaction. If perceptual access to the face of another person is somehow important for the ethical choices we make, then the consequences of losing perceptual access to another agent's face may also become apparent where this loss is due to separation in space and time – and not only where the loss is due to the interaction being mediated by computers. Relevant examples would be the geographical distance between people on each side of the north-south divide or the distance in time between present and future generations. However, since we are focusing in this book on the specific conditions of certain kinds of interaction in cyberspace, we will delve into these implications only to a limited extent.

The Levinasian theme will be given a twist in this book. This has to do with some of the other sources of inspiration encountered in the early days in the writing of this book, primarily the telerobotic art project 'Legal Tender'.

1.1.2 The 'Legal Tender' experiment

The telerobotic art project 'Legal Tender' was a fascinating cyberspatial experiment in which people visiting a web page were presented with a pair of purportedly genuine US$100 bills. After registering, they were offered the opportunity to experiment with the bills by burning or puncturing them online. Having chosen an experiment, the participants were informed that it is a Federal crime knowingly to deface US currency – punishable by up to 6 months' imprisonment – and were left with the option to click the button 'accept responsibility'. Having accepted responsibility, the participants were then allowed to conduct the chosen experiment. In the end the participants were asked if they believed the bills and experiment to be real. Almost all responded negatively.[7]

---

[7] Goldberg in [34, p. 60].

This quite simple but nonetheless immensely original and well-designed experiment seems to provide evidence that something essential to moral life is different when it comes to interaction in cyberspace – a difference that may explain a tendency to act morally differently in cyberspace. This may seem to be a rather imaginative interpretation of the experiment as it is outlined above. It is, however, based on two fairly reasonable assumptions. The first is that, when the participants afterwards responded that they did not believe the bills and experiment to be real, this response reflected a further belief: namely that, had they believed the experiment to be real, they would have acted differently. The second is that the experiment's lack of persuasiveness – or whatever we choose to call it – was somehow constituted by the experiment happening in cyberspace. That is, had the experiment been conducted outside cyberspace, then the participants would not have believed it to be unreal. Although the experiment may point towards a change in the conditions of moral agency in the move from interaction outside to interaction inside cyberspace, no definite conclusion can be reached as it is clear that both of the assumptions vital for the interpretation given here can be rejected. The second assumption, in particular, may seem dubious to some people. Note, however, that the likelihood of people burning or in other ways experimenting with $100 bills if they were given the chance to do so outside cyberspace does not in itself disqualify the assumptions made regarding the experiment. People outside cyberspace may, for example, have reasons to experiment with the bills that overrule or weaken their belief in their genuineness, or they may feel they have relinquished moral responsibility for their experimentation with the bills, for example through being required to do so by an authoritarian scientist overseeing the experiment. In the case of the 'Legal Tender' experiment no such overruling belief appears to be involved: the participants simply stated that they did not believe the experiment to be real, suggesting that this was the reason of their illegal behaviour. In any event, the tendency to act morally differently in cyberspatial or computer-mediated interaction may be further supported.

Some of this support may come from a consideration of the character of conversations in chat-rooms and discussion forums in cyberspace. At the end of 2004 a headline in a leading Danish newspaper declared that 'We sling mud at each other on the internet' – and a follow-up a little more than a year later stated that 'Students are smearing their teachers on the internet'.[8] The content of the first article related directly to the nature of conversation in chat-rooms and discussion forums. A number

---

[8] My translation. Cf. [40, 95].

of researchers as well as representatives of companies providing online chat and discussion reported that the tone and content of conversation on the internet generally
are obscene, disrespectful and offensive. Apparently several of the forums involved had
been closed down because of the nature of conversation online and their inability to
control or change it. On the assumption that the relevant forums were not frequented
exclusively by people of dubious moral character but rather by people with widely
varying backgrounds and moral codes, it seems as though this general trait of conversations in chat-rooms and discussion forums points to a difference between interaction
inside and outside cyberspace. After all, most conversations outside cyberspace are
undertaken in accordance with minimal standards of dignity, respect and politeness.

The last and most informal source of support for the alleged moral difference between interaction inside and outside cyberspace is the numerous conversations I have
had with people frequently engaging in interaction with others in cyberspace. Several
people have testified that they have wondered at the apparent difference in behaviour
they find when they engage in, for example, on-line game-playing with others as compared to face-to-face game-playing. Displeasure with others seems to be expressed
more often and in much stronger language than with face-to-face game-playing with
the very same people. If these testimonies are reliable, then they also seem to support the claim that there is a moral difference between the way people behave inside
and outside cyberspace – a difference that has to do with the fact of the interaction
happening in cyberspace.

Although the examples outlined above do lend some support to the idea that
people act ethically differently inside and outside cyberspace, they do not provide
conclusive evidence to this effect. Nor are they supposed to be seen as doing so. They
can be regarded as sources of inspiration in the sense that they have generated what is
taken as the basic premise of this book, namely that *interaction in cyberspace differs
ethically from interaction outside cyberspace by virtue of agents occasionally acting
in ways in which they would not have acted had they been involved in interaction
outside cyberspace.*[9]

---

[9] A final remark on the sources of inspiration. As stated in the opening of this chapter, the investigations central to this book are also intended to tell part of the story of how spatial and
temporal distance may affect behaviour. That the spatial distance created by new technology
does have ethical implications is most inspiringly thematized in Gunther Anders' book based on
letters exchanged with the pilot Claude Eatherly, who gave the 'go-ahead' for the dropping of
the first atomic bomb after having conducted a reconnaissance flight over Hiroshima. See [2, pp.
1 and 52].

1.1.3   Explaining the basic premise

The combined impetus provided by revisiting Levinas' thinking about 'the face of the other' and by learning about the 'Legal Tender' experiment have been significant in shaping this book.

First, Levinas made it evident that the situation of being face-to-face is ethically relevant. Second, the 'Legal Tender' experiment provided evidence for the claim that there is a moral difference between the way people behave inside and outside cyberspace – a difference that may have to do with the notion of being convinced of the reality of someone or something. In the book these strands come together in an attempt to explain the relationship claimed in the basic premise introduced above, by assessing the evidence required for an agent to be convinced of the reality of the particular interacting agent, but also by weighing up the availability of this evidence in cyberspace. Or, to put it the other way round and in a slightly amended form, this book considers how an agent's belief in the reality of a particular other person may be tied to certain other beliefs, and looks at how the conditions of certain kinds of interaction in cyberspace may limit the availability of the evidence required to form these underlying beliefs. This is followed by a consideration of whether and how the limitations on the available evidence may ultimately come to explain the difference in behaviour captured in the basic premise. By taking these two steps the 'Levinasian gap' is filled. That is, taking these two steps will show how the face of the other is of relevance for the ethical character of interaction, and how the absence of the face of the other may come to influence behaviour in cyberspace.

A few remarks must be made here concerning the overall idea of focusing in this book on explanation. First, it is worth emphasizing the rationale for pursuing an explanation of the basic premise. The focus on explanation is rooted in a desire to show in detail how the change in the setting from outside to inside cyberspace may change the conditions for the formation of certain pivotal beliefs, and how the absence of these beliefs may come to influence behaviour. As will become clear, the explanatory model applied throughout this book allows for a fairly detailed and structured study of each of the elements that are relevant in relating a metaphysical difference to a difference in the behaviour of interacting agents. It is this detailed and structured account that we take to be missing from Levinas' thinking on the ethics of the face of the other.

A second and related point is that, although we are trying to provide an explanation of the relationship claimed in the basic premise, we do not take this to be the only possible explanation. A number of other factors are relevant to the way in which agents behave inside as compared to outside cyberspace. It seems, for instance, that the anonymity of interaction in cyberspace may explain a difference in behaviour. After all, we seem to behave differently on certain occasions if we enjoy anonymity. The difference may also have to do with the fact that in cyberspace information is generally publicly available, whereas in face-to-face interaction information is generally private and ephemeral. This difference may clearly lead to a difference in behaviour. The difference may also have to do with our relationship to new technology. More specifically, it may have to do with the sense of 'alienation' that is often experienced in dealing with new technology. Several other factors could be cited here as relevant for an explanation of the basic premise. However, as has already been made clear, we will in this book focus on the role of an agent's belief concerning the reality of the particular interacting agent. It is a study of the constituents and role of an agent's belief in the reality of the particular interacting party for the agent's behaviour. This narrowing of focus stems in part from the sources of inspiration. Thus the inspiration from Levinas has directed attention towards the ethical relevance of the face of another person, whereas the 'Legal Tender' experiment introduced the relevance of believing something or someone to be real. Moreover, the narrowing of focus reflects the belief that the influence of other factors on behaviour may to some extent be further explained by the framework provided in this book. Thus in the last chapter several alternative explanations are considered, and it is argued that these may in turn come to hinge on the framework provided in this book.

Third and finally, in explaining why people's actions differ according to whether they are operating inside or outside cyberspace we will employ mainly conceptual considerations. That is, the assumed difference in behaviour will be given an explanation that is speculative, in the sense of trading mainly on philosophical concepts and considerations. Thus the book is not based on empirical material in the form of quantitative or qualitative data and statistics collected by means of surveys, experimentation or other kinds of observation. It will, however, exploit empirical material in the sense of incorporating the study of human action in certain manufactured cases on the basis of our intuitions concerning an agent's beliefs and actions in these cases. It will also contain references to empirical studies.

The choice of a speculative approach should come as no surprise, given the aim of suggesting an alternative to Levinas' account of the role played by the face in the ethics of interaction. The principal underlying reason for this choice is simply that the object of study in this book is idealized. As has already been suggested, the book seeks to shed light on the relations between the deliberative and appropriately disposed agents' formation of beliefs and their actions in the context of certain kinds of interaction in cyberspace. The fulfilment of this aim presupposes the ability to exclude from these investigations a number of psychological and sociological factors, for example, which it is notoriously difficult to exclude from any empirical study.

In the light of these considerations one very well may – and should – view the explanation provided in this book as hypothetical in nature. Accordingly, the book may as a whole be seen as the development of a theory or model concerning interaction inside and outside cyberspace. Although the book is limited to the development of a theoretical framework or a model, it is hoped, however, that this theoretical framework or model will be employed in future empirical research reintroducing the complexities removed here.

1.1.4   Road map

A few guiding remarks on the nature and structure of the investigations that follow may be of help in negotiating the way ahead.

Applied philosophy and IT ethics

The work presented in this book lies thematically within the field of moral philosophy and psychology, and the approach taken in dealing with the issues in question draws on a more general analytical framework found in the Anglo-Saxon literature within this field. As a piece of philosophy, the book belongs to the category of applied philosophy. It applies a range of philosophical notions, categories, distinctions, considerations, arguments, assumptions and models in the attempt to illuminate the possible mechanisms underlying the assumed ethical difference between interaction inside and outside cyberspace. In short, it applies philosophy in the attempt to understand a behavioural phenomenon taken from contemporary everyday life.

The book is also intended as a contribution to the demarcation of the new field of IT and computer ethics. One of the predominant debates within this emerging field is the question of the relationship between the ethics of face-to-face interaction and the ethics of interaction involving computers and information technology. At the centre of this debate lies the question of whether this new kind of mediated interaction is accompanied by unique ethical problems and questions. Responses to this question may be and have been attempted from several different angles.[10] This book may be taken to contribute to this discussion by its attempt to defend the idea that the conditions of certain kinds of interaction in cyberspace differ from the conditions of face-to-face interaction, and that this difference may be reflected in a difference in the ethical character of the interaction. Although our considerations throughout the book may also be applied to interaction involving spatially and temporally separated agents, the considerations do point, if not to the ethical uniqueness of certain kinds of interaction in cyberspace, then at least to these kinds of interaction being ethically special.

Structure of book

As the table of contents states, this book is in three parts: 'The basic premise', 'Action, explanation and cyberspace' and 'Explaining the basic premise'.

The first part of the book consists of the first two chapters, the first of which has introduced some of its motivating factors alongside a preliminary formulation of the basic premise of the book. In Chapter 2 the basic premise is further elaborated on, with the purpose of clarifying its precise content. This is clearly of great importance since the remaining parts of the book are aimed at building an explanation of the basic premise.

The second part of the book encompasses Chapters 3 and 4. Chapter 3 contains an inquiry into central concepts of action and explanation: i.e., deliberation, motivation, reasons for action and so on. Although we look at the philosophical discussions surrounding these concepts, the inquiry is conducted in order to construct an explanatory model to be used in the chapters to follow. Chapter 4 is dedicated to an analysis of the concepts of cyberspace, computer-mediated interaction, the internet, virtual interaction, information and so on. The analysis will focus on characterizing those kinds of interaction in cyberspace with which we are concerned in this book.

The third part of the book consists of Chapters 5–9. Chapter 5 is introduced by a section outlining four stages, the last of which is intended to provide an explanation of the basic premise. In the rest of Chapters 5 and 6 the first stage is covered through an exploration and discussion of four hypotheses linking an agent's belief in the reality of

---

[10] See e.g. Birrer in [6], Bynum in [13], Floridi in [26], Gert in [32], Gotterbarn in [37], Gotterbarn and Rogerson in [38], Johnson in [50], Ladd in [54], Maner in [67], Mason in [69], Moor in [71], Moor in [72], Steinhart in [92] and Van den Hoven in [98].

a particular entity with certain other beliefs regarding the properties of that particular entity. Chapter 7 fills in the second stage through an investigation and discussion of the availability in cyberspatial interaction of evidence relevant for the formation and justification of beliefs regarding the interacting party. The third and fourth stages are covered in Chapter 8, in which, first, the role of a belief in the reality of the particular interacting party for an agent's actions is investigated, and second, the threads of the preceding stages are pulled together in an explanation of the assumed difference between interaction inside and outside cyberspace. Chapter 9 contains some concluding remarks.

Reader's guide

Realizing that readers may have different kinds of interest in this book, some guidance on the relative importance of the different chapters may be useful. There are, at least, three possible ways of reading the book. First, it may be read with an interest in the ethics of cyberspace: i.e., with an interest in why people apparently act differently inside and outside cyberspace. In this case the book must be read in its entirety.

Second, the book may be read with a special interest in the considerations on the origins of our belief in the reality of a particular other person, on the relationship between this belief and acting so as to extend moral concern, and finally on how cyberspace may come to influence the formation of these beliefs. In this reading only the later chapters of the book are relevant: Chapters 1–3 and the first two sections of Chapter 4 may be skipped. Weight must then be given to the later section of Chapter 4 and Chapters 5–9.

Third, the book may be read with a very narrow interest in the considerations of those beliefs underlying the belief in the reality of a particular interacting party. In this case, only Chapters 5 and 6 are of relevance.

Chapter 2

The basic premise revisited

Central to this book is the notion of explanation. We will develop a model that will explain the relationship somewhat tentatively developed and denoted in the opening chapter as our basic premise. This premise states that *interaction in cyberspace differs ethically from interaction outside cyberspace by virtue of agents occasionally acting in ways in which they would not have acted had they been involved in interaction outside cyberspace.* It is most unfortunate but perhaps not surprising that the initial formulation of the basic premise suffers from several weaknesses that cannot be ignored. These shortcomings must be more closely examined – and a revised version of the basic premise presented.

T. Ploug, *Ethics in Cyberspace: How Cyberspace May Influence Interpersonal Interaction*,   **13**
© Springer Science+Business Media B.V. 2009

2.1   Shortcomings of the basic premise

2.1.1   The kind of mediation

The basic premise contrasts interaction inside cyberspace with action outside cyberspace. This contrast is, however, rather ambiguous. It may be taken to imply a distinction, on the one hand, between computer-mediated and non-computer-mediated interaction and, on the other, between mediated and unmediated interaction in the sense in which unmediated interaction is face-to-face interaction.[1] The contrast between interaction inside and outside cyberspace runs across these distinctions. Thus it is to be taken as implying a distinction between computer-mediated interaction and unmediated or face-to-face interaction. However, as interaction in cyberspace may be considered to be a species of computer-mediated interaction, in what follows we will contrast interaction in cyberspace with unmediated or face-to-face interaction. Evidently this way of understanding the contrast also has to be built into the basic premise. Note, however, that, aside from the basic premise, we will continue to contrast interaction inside cyberspace with interaction outside cyberspace. Henceforth this contrast is clearly to be read as being between interaction in cyberspace and face-to-face interaction.

2.1.2   The character of actions contrasted

Secondly, the formulation of the basic premise given above is unclear as to the specific character of the actions contrasted.

The moral character of intentional actions

As the interpretation of the examples more or less clearly indicates, however, the difference assumed is between intentional actions and their moral or ethical character. Let us briefly elaborate on these concepts. First, an intentional action is an action undertaken for a reason. This is treated in greater detail in the following chapter, where we deal with the basic concepts of action. Second, the moral or ethical character of an action refers to the properties of an action that make the action either right or wrong according to some ethical theory determining the ethical requirement of a

---

[1] We will not address the question of whether unmediated interaction is at all possible. Hence mediation solely refers to computer mediation.

given situation, where a property is defined as a certain way in which a thing or object is instantiated which may be expressed in a predicative sentence.[2]

Let us illustrate the relevant kind of properties of an action by listing the basic moral requirements of a number of well-known ethical theories, and then relate the relevant properties of an action to these requirements. If, for instance, the moral requirements of a situation are determined by a crude hedonistic utilitarianism , then the relevant properties of an action are those that, taken together, determine whether or not the action maximizes sensual pleasure and minimizes pain for the greatest number of people under the given circumstances.[3] If the moral requirements of a situation are determined by a Kantian ethic of duty, then the relevant properties of an action are those that, taken together, determine whether or not the action is in accordance with the duties as derived from one of the formulations of the categorical imperative.[4] If the moral requirements of a situation are determined by an Aristotelian virtue ethic, then the relevant properties of an action are those that, taken together, determine whether or not the action is a mean between excessive actions: i.e., between doing too much and doing too little.[5]

Although this may seem obvious, it is worth noting that two actions differing in their relevant properties do not necessarily differ in their moral status. That is, they do not necessarily differ in being morally right or wrong. For this to be the case the situations would have to be identical in terms of their moral requirement: i.e., the same in terms of the properties constituting their moral requirement. Second, there would have to be only one action characterized by a unique set of properties satisfying the moral requirement of the situations. It follows, therefore, that the basic premise is not a claim to the effect that the difference between interaction inside and outside cyberspace is a difference in the moral status of the actions – i.e., it is not a claim to the effect that an action within the one domain is morally right whereas its counterpart in the other domain is morally wrong. For all we know, it could be that the situations are not identical in the properties constituting the moral requirements, and actions could

---

[2] Cf. Martin in [68] and McNaughton in [70, p. 193]. The concept of property applied here may be considered controversial. However, the metaphysical discussion of properties is taken to be neutral in relation to the considerations of this book. That is, it may at most influence the way in which the basic premise is to be formulated.

[3] Cf. Bentham in [5, pp. 14–15].

[4] Cf. Kant in [52, pp. 68ff. and 79].

[5] Cf. Aristotle in [3, 1106a and 1106b].

thus differ in their moral properties without differing in their moral status. For this reason the basic premise must be limited to a claim regarding a difference between the moral properties of otherwise comparable actions. This is discussed further in the subsections below on the third and fourth shortcomings of the initial formulation.

Virtual and counterfactual actions

As regards the character of the actions contrasted, the basic premise does make an important feature of these actions explicit. Thus the basic premise states that agents act differently in cyberspace from how they would have acted outside cyberspace. Hence the basic premise is not contrasting two actual actions but rather an actual action with a counterfactual one. Counterfactuals are usually taken to be non-truth-functional conditionals: i.e., conditionals whose antecedents are false without this implying the truth of the conditional.[6] An example would be: 'If the computer had not been invented, this book would not have been written.' As is evident from this example, counterfactuals enable a contrasting of actions or events in situations that differ from each other in a limited and specifiable way. In other words, counterfactuals enable the contrasting of an action in an actual situation with an action in a similar, but imagined or construed, situation. In short, then, a counterfactual action is an action in a non-actual, possible world given as the consequent of a non-truth-functional conditional.

2.1.3   Coincidental difference in interaction

The basic premise states that occasionally agents act differently in cyberspace from how they act outside cyberspace. This formulation is rather problematic, since it leaves it undecided whether or not the assumed difference in behaviour occurs entirely as a matter of coincidence.

The difference in behaviour may be coincidental or accidental in at least two ways. For example, this will be the case if the difference in behaviour reflects the fact that the interacting agents occasionally and for no good reason decide their course of action on the basis of tossing a coin. In other words, it will be the case if, for no good reason, either a random procedure for deciding a course of action or a random procedure for deciding the procedure for deciding a course of action is applied. In a generalized form

---

[6] Cf. [47, p. 169].

it will be the case if, for no good reason, a random procedure for decision-making is applied at any level of higher- and lower-order procedures of decision-making. On the other hand, it will also be the case if the difference in behaviour reflects the fact that the interacting agents are occasionally unable to act in accordance with the decided course of action because of randomly suffering weakness of the will, accidie, tiredness, illness, apathy, a feeling of uselessness or some other form of so-called 'depression'.[7] In other words, it will be the case if depression strikes randomly.

Acting differently because of a difference

To assume that the difference in interaction simply reflects the interacting agents applying a random procedure of decision-making or reflects the agents randomly suffering from a depression is clearly not intended by the basic premise. If that were the intention, the basic premise would not have contrasted interaction inside cyberspace with interaction outside cyberspace. Rather, it would have contrasted any two situations with a difference in the actions of the interacting agents, since any such difference in behaviour may reflect either a random procedure of decision-making or a random suffering of depression. The basic premise obviously contrasts interaction inside and outside cyberspace because the difference in interaction is assumed to reflect a difference between the situations in which the agents are to decide their course of action and subsequently act. (Note that the situations are here assumed to be describable independently of the difference in behaviour.) That is, there is here a difference between the conditions or circumstances to which the agents are to apply a procedure of deciding a course of action and subsequently to implement this choice of action. Amending the basic premise so as to claim that the difference between behaviour inside and outside cyberspace reflects a difference between the relevant situations clearly excludes from its scope differences in behaviour reflecting a random application of a random procedure of deciding a course of action or a random suffering of depression. Unfortunately it does not, however, exclude the possibility of the difference in behaviour occurring as a matter of coincidence.

Again this may come about in at least two ways. For example, a difference in behaviour will be coincidental if the difference in behaviour reflects the fact that the interacting agents, for no good reason, decide their course of action by means of tossing a coin whenever the interaction is computer-mediated and they therefore have

---

[7] Stocker in [94, pp. 744–745] and Smith in [87, p. 120].

limited perceptual access to each other. In other words, it will be the case if, for no
good reason, a random procedure for deciding a course of action is applied, although
the decision as to when to apply the procedure is not random – that is, although no
random higher-order decision-making procedure is at work. On the other hand, it will
be the case if the difference in behaviour reflects the fact that the interacting agents
suffer depression whenever the interaction is computer-mediated and they therefore
have limited perceptual access to each other. In other words, it will be the case if the
suffering of depression is somehow related to contingent features of the world.

Note, crucially, that in both these examples the difference in behaviour reflects a
difference between the situations. However, the difference is still coincidental or, at
least, contingent. In the first example it is coincidental in the sense employed hitherto.
Thus the application of the random procedure of decision-making results in a random
choice of action for no good reason. In the second example, however, the difference in
behaviour is obviously not coincidental in the sense of occurring as a matter of random
chance. Rather, there is a non-random correlation between the difference in situations
and the difference in behaviour. The difference is still coincidental – although now in
the significantly weaker sense according to which it is merely a contingent psychological fact about the interacting agents that eventually results in the difference in
behaviour. In other words, the difference in behaviour merely reflects how the world
and the people in it happen to be equipped or functioning. To summarize: in order to
rule out entirely the possibility of the difference in behaviour reflecting coincidences
or contingencies, it is not sufficient to amend the basic premise to include the claim
that the difference in behaviour reflects a difference between the situations. However,
this amendment clearly takes us some of the way by excluding the possibility that the
difference in interaction reflects pure coincidence, as was illustrated in the first two
examples of this subsection.

Acting for a strong normative reason and fully disposed

Hardly surprisingly, the basic premise is intended to rule out the possibility of the
difference in behaviour reflecting coincidences or contingencies as just exemplified.
This would be achieved if the basic premise were rephrased so as to incorporate the
following claim: there is a moral difference between behaviour inside and outside
cyberspace, which occurs not because agents for no good reason apply a random
procedure in deciding their course of action or because agents suffer from depression,
but rather because of a difference between the two situations.

At this point let us pause for a moment to inquire into the possibility of positively describing the character of the agents who act differently in cyberspace without this difference in behaviour reflecting either a random procedure of decision-making or the effect of depression. In terms of the latter, the task is quite easy. The agent not suffering from depression is, defined positively, the agent who is psychologically fully disposed to act in accordance with the moral requirements of a situation. In terms of the former, the task is only a little more difficult. Positively defined, the agent who does not, for any good reason, apply a random procedure for deciding a course of action is obviously the agent who applies a random procedure for decision-making only if there is a good reason to do so. In order to increase the non-tautological content of this definition, let us tentatively say that a good reason is a strong normative reason for action, where a strong normative reason is a reason implied by a normative theory satisfying some minimal requirements of excellence, such as logical consistency and coherence. Thus such agents apply a random procedure for decision-making only if they endorse a logically consistent and coherent normative theory of action that prescribes the application of a random procedure, such as tossing a coin, for deciding the course of action.

Acting for a strong normative reason or because of a difference

An important question presents itself here. Suppose the basic premise is rephrased as suggested above. It will then have two components. First, it will claim that agents acting for a strong moral reason and fully disposed to act in accordance with the moral requirements of situations act differently inside and outside cyberspace. Second, it will claim that the difference in interaction reflects a difference between the situations. The question that inevitably arises is what the logical relationship between these components is. Since we have already seen that only the first component rules out certain kinds of coincidence, we may infer that the second component does not imply the first. It thus remains to be considered whether the first component implies the second. Thus it may be that the agent acting for strong normative reasons will act differently in two situations only if there is a relevant difference between the situations. That is to say, it may be that acting for normative reasons implies that agents only act differently in two situations if there is a difference between them. Now acting for a normative reason does not exclude the possibility of acting differently without there being a difference between the situations. If one holds inconsistent and incoherent beliefs, then one may, according to standard logical calculus, draw different conclusions as to what one is supposed to do and consequently act differently in two relevantly identical situations. Acting for a strong normative reason, however, rules this possibility out if we follow the tentative definition given above, as it implies acting on reasons provided by a consistent and coherent normative theory.

A possibility of acting for a strong normative reason but nevertheless differently in two situations that are relevantly identical arises owing to the possibility of committing epistemological mistakes such as misperceiving, misinterpreting, misjudging or misrepresenting the properties of a situation. If it is possible to form false beliefs – i.e., beliefs not corresponding to facts – in one of two relevantly identical situations by committing any one of these mistakes, and if it is possible to make such mistakes in one of the situations without this ultimately reflecting a difference in the properties of a situation, then it is also possible that an agent acting for a normative reason may act differently in two situations without this reflecting a difference between the situations. If these possibilities are allowed, it seems as though we cannot but conclude that agents acting for a strong normative reason do not only act differently if there is a difference between the situations. Hence the first component does not logically entail the second.

Epistemological coincidences and contingencies

Apart from serving the purpose of being counter-examples to the suggested amendments of the basic premise, the examples above also serve another important purpose. They illustrate another way in which the difference between behaviour inside and outside cyberspace can be coincidental or accidental. Thus epistemological failures leading to a difference in behaviour in situations with no relevant difference would be coincidental exactly in the sense that they strike at random. Bearing in mind this possibility as another kind of coincidental difference in behaviour, we have to consider yet again whether the basic premise is intended as making a claim regarding such differences in behaviour. Given our previous considerations, the answer seems evident. Again, the basic premise would not specifically contrast interaction inside and outside cyberspace if it was intended as a claim to the effect that the difference in interaction reflects agents randomly making epistemological mistakes. Hence we have yet another reason for maintaining, as part of the basic premise, the claim that the difference in interaction reflects a difference between the situations of interaction. This would obviously rule out purely random epistemological mistakes.

The parallels to the kinds of coincidental behaviour discussed throughout the previous subsections of this section clearly do not end here. Once again, the claim that the difference in interaction reflects a difference between the situations does not exclude the possibility of other kinds of epistemological failures causing the difference to be contingent. We could speculate whether, owing to some kind of contingent psychological mechanism, agents are more likely to commit epistemological failures such as misperceiving, misinterpreting, misjudging or misrepresenting the properties of a situation whenever interaction is mediated by a computer screen. In contrast to the case of people suffering depression because of the computer mediation, however, the basic premise is not intended to rule out such cases of contingent, psychological mechanisms resulting in epistemological mistakes and, ultimately, in the assumed difference in behaviour, as long as they may be adequately explicable on the basis of conceptual and *a priori* investigation: i.e., without reference to *a posteriori* studies of human psychology.

Hence it is tacitly assumed that the difference between the psychology of so-called depression and the psychology of epistemology lies in the former requiring extra-philosophical means to be described adequately. Although these considerations may be thought to be highly controversial, we will not pursue further justification of the position adopted. It is simply the view of this book that, at least occasionally, the psychology of forming, justifying and acting on beliefs may be adequately described as a result of conceptual studies. For this very reason we will not hesitate in later sections to make use of such epistemological concepts in our explanation of human behaviour.

To bring these considerations of coincidental and contingent epistemological failures to a conclusion, let us simply emphasize that, in order to avoid undesirable forms of epistemological coincidence, it is sufficient to maintain the second component of the basic premise: that is, the claim that the difference between interaction inside and outside cyberspace reflects a difference between the situations. This fits nicely with the conclusion above: namely, that agents acting for a normative reason and being adequately disposed do not only act differently in the two situations if there is a difference between the situations. Hence there is also a need in the basic premise for the explicit claim that agents act differently because of a difference.

Before moving on to consider a fourth shortcoming, it is worth noting that the investigation of this subsection has, in line with the preceding subsections, further narrowed down the character of the agents and actions to be focused on in the chapters that follow. In the previous subsection we concluded that the basic premise concerned the moral character of intentional actions. From the investigations of this subsection we may now conclude that it concerns the moral character of actions done for a strong normative reason by agents suitably disposed and changing their course of action because of a change in the properties of the situation. With this in mind let us consider a fourth shortcoming of the preliminary formulation of the basic premise.

2.1.4   Qualitative identity of situations

The tentative formulation of the basic premise obviously compares a situation of interaction inside cyberspace with a hypothetical situation of interaction outside cyberspace. On the basis of this comparison comes the assertion that agents intentionally act differently in the two situations. In the absence of coincidental behaviour, a comparison between situations is normally carried out in order to indicate that a certain difference or similarity in the properties of the situations is reflected in a certain difference or similarity in behaviour. In the absence of coincidental behaviour, a comparison is thus indicative of a limited number of properties that are either sufficient for the occurrence of a difference in behaviour in situations sharing all other properties or sufficient for the occurrence of a similarity in behaviour in situations sharing only these properties.

Comparing situations

Imagine, for instance, that an agent in a situation S_1 characterized by the sets of properties P_1 and P_2 does A, and the very same agent in a situation S_2 characterized by the sets of properties P_1 and P_3 does $\neg A$. Given that $\forall p_x \forall p_y (((p_x \in P_2) \wedge (p_y \in P_3)) \rightarrow (p_x \neq p_y))$, then a comparison between the situations would be indicative of a limited set of properties sufficient for the occurrence of the difference in behaviour. Thus it may inferred that the agent acted differently in situation S_2 from S_1, because of S_2 was characterized by the set of properties P_3. This holds although both sets of properties in both situations may be claimed to be sufficient for the production of action. The claim advanced here concerns the set of properties sufficient for a difference in behaviour to occur – not the sets of properties sufficient for action. If the agent in the example just given had done A in both situations, S_1 and S_2, then the comparison would have been indicative of a limited set of properties sufficient for the occurrence of this similarity in behaviour. The agent acted the same way in situation S_2 as in S_1, because of S_2 being characterized by the set of properties P_1.

In this context the lesson to be learned from these considerations is this: a comparison of behaviour involving two completely different situations cannot be indicative of a limited number of properties sufficient for the occurrence of a difference or similarity in behaviour between the situations. Hence, if the basic premise is intended to contain such an indication, it must tacitly assume that the situation of interaction inside cyberspace is not wholly different from the hypothetical situation of interaction outside cyberspace. Or to put it slightly differently, it must tacitly assume a certain degree of similarity or sameness to hold between the two situations. The task now is to decide whether the basic premise is intended to contain such an indication, and if so, how the tacitly assumed similarity or sameness can be precisely described. The result can thus be fitted into the revision of the basic premise as an explicit claim about the identity of the situations of comparison.

The first task is readily completed. Since the basic premise specifically contrasts interaction inside with interaction outside cyberspace, it is clearly intended to be indicative of a set (a category) of properties sufficient for the occurrence of the difference in behaviour between these two kinds of interaction, namely a subset of the set of properties distinguishing acting inside cyberspace from acting outside cyberspace: that is, a subset of the set of new properties entailed by the property of the interaction taking place inside cyberspace or being computer-mediated.

The second task is then to describe precisely the similarity or identity tacitly assumed to hold between the two situations of interaction and then fit it into the basic premise as an explicit claim about the identity of the situations being compared. To do this, however, we need to overcome several problems. In the first place, qualitative identity may refer to different kinds of properties that are not necessarily co-extensional. In the second – and this is closely related – demanding qualitative identity in terms of a kind of property may be a requirement that cannot be satisfied. Let us elaborate on each of these problems in turn.

Essential and morally relevant properties

Let us assume that any given entity – and therefore any given situation – may be fully described by a set of essential and non-essential properties. The essential properties are those defining characteristics encapsulating the very nature of an entity. A property may, however, be essential in more than one sense. Of immediate relevance for our purposes is the idea that a property may be essential in the sense that it is an essential property for every entity possessing it: i.e., if an entity possesses this property, then

it necessarily possesses it as an essential property. By contrast, a property may also
be essential in the sense that it may be essential to one entity possessing it but not
essential to another entity possessing it.[8]

Let us further assume that a situation may be fully described by a set of morally
relevant and irrelevant properties, where the morally relevant properties are the properties relevant in determining the moral requirements of the situation. The morally
relevant properties may be constituted by moral or non-moral facts about the agent
and the context. Note, very importantly, that the morally relevant properties are
those kinds of properties that are pointed out by the application of normative ethical theories or principles – that is, the kind of properties selected by, for instance,
applying an ethical principle such as the basic utilitarian principle, the categorical
imperative or the Aristotelian principle of the virtuous action as a mean between
doing too much and doing too little. The distinctions between the set of essential
and non-essential properties and the set of morally relevant and morally irrelevant
properties may be applied to the same set of properties. However, they are not necessarily co-extensional. Thus the essential properties of the situation may be a subset
either of the morally relevant properties of the morally irrelevant properties or of the
intersection of the two, and vice versa. Ultimately, then, two distinctions between
kinds of properties suggest themselves as candidates for describing the qualitative
identity assumed to hold between the situation of interaction inside cyberspace and
the situation of interaction outside cyberspace.

To describe the qualitative identity tacitly assumed by means of any of the kinds
of properties defined above seems quite difficult, since it is not at all clear whether
it is an essential, non-essential, morally relevant or morally irrelevant property of
interaction in cyberspace that it occurs in cyberspace. In fact, the property of occurring in cyberspace may entail a set of genuinely new properties of the situation of
interaction, some of which are essential, others non-essential – some morally relevant,
others morally irrelevant. Moreover, if the set of new properties cannot readily be
fitted into the relevant kinds of properties, then there is a risk of rendering the basic
premise inconsistent by describing the qualitative identity tacitly assumed to hold
between the situations of interaction in terms of any of the kinds of properties. Thus
it may be precisely that set of properties entailed by the property of the interaction
being in cyberspace that is reflected in the difference in behaviour. If, therefore, by

---

[8] Cf. [47, pp. 250–251].

means of a comparison between two situations the basic premise is supposed to be indicative of a set of properties that distinguishes the two situations and is sufficient for a difference in behaviour to occur, then obviously it would be inconsistent for the basic premise tacitly to require the situations to be qualitatively identical in terms of the relevant set of properties. These considerations also rule out the possibility of describing the qualitative identity tacitly assumed in terms of any combinations of kinds of properties.

More importantly, even if it could be settled that none of the new properties entailed by the property of the interaction happening online figured among the one kind of properties by means of which the qualitative identity may have been described, then the occurrence of a difference in behaviour between two situations being qualitatively identical in the relevant way would not be sufficient to make the difference in behaviour precisely reflect some or all of the members of the set of new properties entailed by the interaction happening in cyberspace. In short, the basic premise would not necessarily indicate that some or all of the members of the set of new properties entailed by the property of the interaction happening inside cyberspace or being computer-mediated were sufficient for the difference in interaction to occur. If, for example, the qualitative identity only concerns essential properties, then it is logically possible that the two situations do not share any non-essential properties, in which case the difference in behaviour may reflect non-essential differences other than a non-essential difference consisting of some or all of the members of the set of new properties entailed by the interaction happening in cyberspace. If the qualitative identity concerns only morally relevant properties, the same argument applies. Thus it is logically possible that the two situations may not share any morally irrelevant properties, in which case the difference in behaviour may reflect morally irrelevant differences other than a morally irrelevant difference consisting in some or all of the members of the set of new properties entailed by the interaction happening in cyberspace. (The same argument obviously applies to qualitative identity described in terms of non-essential or morally irrelevant properties or any combination of kinds of properties not making up the total set of properties.)

A moral or non-moral difference

Although it may appear to be discouraging, this attempt at positively describing the qualitative identity tacitly assumed by means of the distinctions between essential and non-essential and between morally relevant and morally irrelevant properties has served an important purpose. Our considerations have indirectly made it clear that the difference between the actions constituting interaction inside and outside cyberspace may possibly reflect both morally relevant and morally irrelevant properties of interaction in cyberspace. In trying to describe the qualitative identity tacitly assumed to hold between the situations of interaction, it has been left an entirely open question whether the subset of the set of properties sufficient for the difference in interaction to occur is a set of morally relevant or morally irrelevant properties. In this way the basic premise is not biased in terms of the kind of property sufficient for the difference in interaction to occur: it may be a difference in the set of properties determining the moral requirement of the situations, or it may be a difference in the set of morally irrelevant properties.

In terms of making an explicit claim about the character of identity of the situations compared to be incorporated into the basic premise, it is now evident that such a claim must be stated in the negative. That is, the claim can only state that the situations compared are identical in that they share all morally relevant and irrelevant, essential and non-essential properties except possibly those that are genuinely new in the sense of being tied exclusively to interaction in cyberspace.

2.2 The basic premise

2.2.1 Restating the basic premise

The considerations of the previous section have indicated the need for a revision to the basic premise as it was stated in the opening chapter.

Before revising the basic premise in the light of the shortcomings pointed out, let us briefly restate it in plain language. The basic premise is supposed to be a claim to the effect that if one compares the moral character of the actions of one or more agents engaged in cyberspatial interaction with the actions of the same agents in a similar, hypothetical situation of interaction outside cyberspace, then there is a difference between the properties constituting the moral character of the actions in the two situations.

Combining this plain statement with the refinements suggested in the preceding section, we may rephrase the basic premise thus:

[TBP]   Interaction in cyberspace occasionally differs from face-to-face interaction by virtue of being constituted by agents performing acts differing in their moral properties. This applies to cases in which: (a) the interacting agents decide their course of action on the basis of a correct deliberation of their strong moral reasons for action and are disposed to act in accordance with the moral requirements of both situations; (b) an agent's behaviour reflects the properties of situations such that a difference between the actions in two situations entails a difference between the properties of the situations; and (c) the situations compared are identical in all properties except in the set of properties distinguishing acting inside cyberspace from acting outside cyberspace – i.e., in the set of properties distinguishing computer-mediated interaction from face-to-face interaction.

Let us briefly examine this phrasing of the basic premise in the light of the shortcomings that have been pinpointed.

The first shortcoming is accommodated by the basic premise ([TBP]) explicitly contrasting interaction in cyberspace with face-to-face interaction. The second shortcoming is accommodated by [TBP] contrasting the moral properties of intentional actions. The third is accommodated by the first two amended claims – i.e., claims (a) and (b) – ensuring that the difference between the agents' behaviour in the two situations of interaction is non-coincidental, non-contingent and reflects a difference between the situations. The fourth shortcoming is countered by the third amended claim, claim (c), ensuring that the situations compared are as similar as possible.

2.2.2   Exploration of the basic premise

The restated version of the basic premise has some noteworthy features, which we will briefly consider in order to avoid possible misunderstandings.

Note, first, that the basic premise does not distinguish between different kinds of interaction in cyberspace. As it is worded in [TBP], it may apply to any kind of interaction in cyberspace: e-mailing, gaming, chatting, e-learning, e-banking, downloading software, making reservations, searching for information, exploring cyberworlds by the use of avatars and so on. We do not, however, take the basic premise to hold for all kinds of interaction in cyberspace. Hence in Chapter 4 we will narrow the focus of our endeavour so as to attempt the explanation of [TBP] for two particular kinds of interaction in cyberspace: interaction in chat-rooms and tele-operation.

Note, second, that even though [TBP] contrasts interaction inside and outside cyberspace, this is not a contrast between interacting entirely inside and outside cyberspace. Thus in its current formulation [TBP] does not imply that the moral difference only shows in the case of agents having entirely substituted interaction in cyberspace for interaction outside cyberspace. In our attempt to explain [TBP] as applied to particular kinds of interaction in cyberspace we will therefore be focusing on providing an explanation that does not presuppose that the interacting agents have substituted to any significant degree life outside cyberspace with life inside cyberspace – but rather explains [TBP] without making any assumptions in this respect.

Note also, third, that [TBP] is independent of any particular normative ethical position. That is, it does not rest on an assumption to the effect that the difference between interaction inside and outside cyberspace only shows in the case of the interacting agents subscribing to particular normative ethical views such as utilitarianism, virtue ethics or any other normative criteria of ethical rightness and wrongness. All it really claims is that there is a difference between the moral properties of actions performed in interaction inside and outside cyberspace by one and the same agent holding the same set of normative ethical beliefs. A fourth and related point to note is that [TBP] does not entail a specific meta-ethical position in terms of the debate between moral realists and anti-realists. That is, it entails neither the acceptance nor the denial of objective, truth-making moral facts. Again, all the basic premise entails is that agents may hold certain moral or ethical beliefs that may come to influence their actions. Whether or not these beliefs are true or false, and thus may be claimed to correspond to moral facts, is simply irrelevant for the explanation of [TBP].

Part II

Action, explanation and cyberspace

Chapter 3

Actions and explanations

The task this book has set itself is to develop a model that will explain the basic premise ([TBP]) by referring to an agent's beliefs concerning the reality of the particular, interacting party. The first step in fulfilling this task was taken in the previous chapter, where the relationship to be explained was elaborated on the basis of a number of shortcomings.

Now, to provide an explanation of the difference in interaction captured in the revised version of the basic premise clearly requires a picture of what we might call the 'mechanics' of interaction in cyberspace. That is, it requires some sort of understanding of the basic elements constituting interaction, of the notion of cyberspace and of how these elements are to be fitted together in order to constitute an explanation of the relationship in question.

The aim of this chapter is to outline different ways of modelling human intentional behaviour as it may be instantiated in interaction both inside and outside cyberspace, using a study of recent works on the concept of moral motivation. In addition to providing an understanding of the elements of human intentional action, these considerations will be assembled into an explanatory model to be used in the explanation of the basic premise ([TBP]) in later chapters.

T. Ploug, *Ethics in Cyberspace: How Cyberspace May Influence Interpersonal Interaction*, **33**
© Springer Science+Business Media B.V. 2009

3.1   Actions and reasons

3.1.1   'The moral problem'

To explain the basic premise ([TBP]) is in part to explain why people act in a certain
way in cyberspatial interaction. Before going any further we must therefore investigate
the notion of an action. This will eventually lead to an investigation of different
kinds of reasons for action, and finally to different possible approaches to the task
of explaining human behaviour. These considerations are included in the attempt
to develop a model for the explanation of the basic assumption ([TBP]), but also,
and very importantly, because they will show how the explanation provided here is
logically independent of the various theories or models of moral motivation. This
clearly adds to the credibility of the conclusions drawn.

Defining action

Let us define an action as constituted by an agent doing something intentionally
through the movement of his or her body. So defined, an action denotes the happening
or occurrence of something; i.e., an action becomes a particular kind of event. Events
are, however, particulars describable in several ways. The event of pointing a finger at
a blackboard may also be described as teaching. For our purposes it suffices that, for
this event to qualify as action, one of the possible descriptions of the event involves a
person doing something intentionally.[1]

The definition of an action on the basis of bodily movement and intentionality is
not wholly unproblematic. Thus it seems as though one may do something intentionally that does not involve bodily movement, such as refraining from helping an injured
person, which nonetheless qualifies as an action in the sense in which an action is doing something for which one may be held morally accountable. Actually, taking this a
step further, it seems as though one may even perform actions unintentionally. Thus
it seems as one may do something unintentionally – e.g., killing a fish by jumping
into the sea – which, again, nonetheless qualifies as an action in the sense in which
an action is doing something one for which may be held morally accountable.[2]

Since we are not here concerned with making any claims regarding the adequacy
of the above definition of an action, and since this definition will suffice for the present

---

[1] Cf. [47, pp. 4–5 and 253–254].

[2] Cf. [47, pp. 4–5].

purposes, we will, however, leave the discussion of possible definitions of an action.
From this point onwards, then, the focus of our investigation will be the notion of
reasons for action.

Normative and motivating reasons

The task of defining the concept of acting intentionally is intimately linked to that of
determining what constitutes a reason for action, since it may be claimed that to act
intentionally involves acting for a reason.[3] Let us distinguish two kinds of reasons for
action: Explanatory or motivating reasons and justifying or normative reasons. Often applied in everyday analysis of action, this distinction reflects, on the one hand,
the reasons that explain why someone acted as he or she did, and, on the other, the
reasons that may or may not justify the relevant course of action. More specifically,
for an agent to have an explanatory or motivating reason to act in a certain way is
for the agent to be in a certain psychological state that – other things being equal –
motivates or produces the relevant action, and thus is explanatory of the agent acting in the relevant way.[4] On the other hand, for an agent to have a justifying or
normative reason to act in a certain way is for there to be a normative requirement
generated by a normative system that the agent acts in the relevant way. Thus the
agent's action is justified from the perspective of the normative system that generated
the requirement.[5] The two kinds of reason are clearly different and independent categories. They are different categories, since motivating reasons are psychological states
whereas normative reasons are requirements best thought of as propositions stating
what is desirable or required by an agent. And they are independent categories by
virtue of the possibility of an agent having motivating reasons without having normative reasons, and vice versa.[6] Henceforth we will refer to the two kinds of reasons
as motivating and normative reasons.

---

[3] Davidson in [18, p. 6].

[4] Smith in [87, pp. 95–96]. Cf. also Brink in [10, pp. 39–40].

[5] Smith in [87, pp. 95–96]. According to Smith there are normative reasons of rationality, prudence, morality etc. Cf. also Brink in [10, pp. 39–40].

[6] Smith in [87, p. 96] and Brink in [10, p. 40].

Internalism versus externalism

An extensively discussed topic within the field of meta-ethics revolves around the question of whether one should be internalist or externalist with regard to moral considerations. In some versions internalism and externalism are opposed by virtue of taking different views as to the possibility of moral requirements or judgements providing motivating or normative reasons for action. Following a well-known exposition, let us introduce a distinction between agent and appraiser internalism.[7] Agent internalism holds that necessarily, if a person ought to do p, then she has motivating or normative reasons to do p. Appraiser internalism holds that necessarily, if a person judges that he or she ought to do p, then she has motivating or normative reasons to do p. Agent and appraiser internalism may be further subdivided by distinguishing between weak and strong internalism. Weak internalism claims that necessarily, if a person ought to do p, then she has some motivating or normative reasons for action. Strong internalism claims that necessarily, if a person ought to do p, then she has sufficient motivating or normative reasons to do p. Henceforth we will leave this distinction aside and solely focus upon the distinction between agent and appraiser internalism.[8]

It is easily seen that agent internalism about motivating reasons is not a tenable position, as it would imply that a person could be motivated to act without knowing it. Likewise appraiser internalism in regard to normative reasons is not a tenable position, since a person may be mistaken about what he or she ought to do; in other words, to judge that something is required does not necessarily provide normative reasons for action. We end up, therefore, with two tenable internalistic positions: agent internalism about normative reasons and appraiser internalism about motivating reasons. Consequently externalism may be construed as a denial of either of these positions.[9] In the following we will for the sake of brevity and relevance focus exclusively on appraiser internalism.

---

[7] Brink in [10, pp. 40–41].

[8] Brink in [10, pp. 41–42].

[9] Brink in [10, p. 42] construes internalism as a conjunction of three claims. Agent and appraiser internalism express only the first of these. Hence in Brink's terminology there are versions of externalism other than those presented here.

'The Moral Problem'

Of relevance for our purpose of developing a model for the explanation of human behaviour are now 'the moral problem' and suggested solutions to this problem. 'The moral problem' simply consists in the incompatibility of three intuitively plausible positions: appraiser internalism, cognitivism about morality and a Humean theory of motivation.[10] Cognitivism is the view that moral judgements express a person's belief about an objective matter of fact as to what it is right for her to do. The Humean theory of motivation holds that a person is motivated to act in a certain way just by virtue of having an appropriate desire and a means-end belief, where belief and desire are distinct existences in the sense that an agent can have one without the other.[11]

As mentioned above, the combination of appraiser internalism about motivating reasons, cognitivism and the Humean theory of motivation is inconsistent. Thus, if moral judgements motivate (appraiser internalism), and moral judgements express a belief about an objective matter of fact (cognitivism), then to be motivated simply cannot presuppose a distinct existing desire.

There seem to be two readings of 'the moral problem'. Either the problem arises because the combination of appraiser internalism and cognitivism implies that motivation is purely cognitive, which the Humean theory of motivation obviously denies,[12] or it arises because the combination of appraiser internalism and cognitivism implies the existence of a necessary connection between believing something and being motivated: i.e., that beliefs may play the main causal role in the production of action. This is something the Humean theory of motivation denies by claiming that belief and desire are distinct existences and that both are required in order for a person to be motivated to act.[13]

3.1.2 A Humean theory of motivation

One position on the presented problem would be to give up either appraiser internalism or cognitivism and maintain a Humean theory of motivation.

---

[10] 'The moral problem' is developed in McNaughton [70, pp. 22–23] and in Smith [87, pp. 11–13 and 125–129].

[11] Smith in [87, pp. 12 and 119].

[12] McNaughton in [70, p. 23].

[13] Smith in [87, pp. 126 and 181].

The combination of appraiser internalism and the Humean theory of motivation implies that moral judgements certainly motivate but that they express desires and the means-end beliefs of agents. Hence moral motivation is not purely cognitive and beliefs cannot play the main causal role in the production of action, as would be implied by the combination of appraiser internalism and cognitivism. The combination of cognitivism and the Humean theory of motivation implies that moral judgements are purely cognitive but do not motivate, as this requires a desire and means-end beliefs. Hence, again, moral motivation is not purely cognitive and beliefs cannot play the main causal role in the production of action, as would be implied by the combination of appraiser internalism and cognitivism.

The quasi-hydraulic view of motivation

The Humean theory claims that a person is motivated to act in a certain way simply by having an appropriate desire and a means-end belief, where belief and desire are distinct existences in the sense that an agent can have one without the other.

The belief–desire theory of motivation is often interpreted by invoking a metaphor of hydraulics. The desire provides the motivational push that moves an agent to act, and the beliefs provide the information as to how to satisfy the desire, thereby channelling the push of the desire in the right direction.[14] In this reading the desire clearly becomes a causal source of action, and thus a hydraulic view for explaining action is one form of causal explanation of human action.

Interwoven with the hydraulic reading of the Humean theory of motivation is the disputed concept of a desire. Thus, following Hume, the traditional interpretation of a desire has been what has recently been dubbed a 'strong phenomenological concept'.[15] In this interpretation a desire is essentially a state with a phenomenological content – a psychological feeling analogous to bodily sensations.

Both the causal nature of the hydraulic reading of the Humean theory of motivation and the strong phenomenological conception of desire have been criticized.[16] For the present purposes, however, this debate is irrelevant.

---

[14] McNaughton in [70, p. 21] and Smith in [87, pp. 101–102].

[15] Smith in [87, p. 105] for the notion of 'strong phenomenological concept'. In [87, pp. 112–113] Smith argues that Hume did not believe desires to have phenomenological content essentially.

[16] Smith in [87, pp. 105–107 and 114–115] and Platts in [78, p. 76].

3.1.3 Internalism in relation to normative reasons

The second approach to the moral problem to be considered here tries to reconcile the Humean theory of motivation, appraiser internalism and cognitivism. It is developed from considerations regarding explanations of intentional actions by means of motivating and normative reasons. The details of this approach are worth a closer look.

Intentional and deliberative explanations

Intentional actions may be explained from both an intentional and a deliberative perspective. From the intentional perspective explaining is done by giving motivating reasons for action. Assuming the Humean theory of motivation to hold for motivating reasons,[17] an example of explanation from the intentional perspective would be: An agent did q (jumped into the sea) because she desired to p (get clean), and had a belief that, were she to p (get clean), she would do so by doing q (jumping into the sea). The belief and desire are psychological states that teleologically, perhaps even causally, explain the action.[18]

From the deliberative perspective explaining is done by giving normative reasons – i.e., propositions that from an agent's point of view are taken to give a rational justification of a given course of action: e.g., an agent did q (jumped into the sea) because, after having deliberated, she thought it to be desirable or required that she p (got clean), and had a belief that were she to p (get clean) she would do so by doing q (jumping into the sea). It is of crucial importance to note that explanation from the deliberative perspective is indeed a way of explaining intentional behaviour. It is, in fact, the case that our reflections regarding what we have normative reasons to do can, at least at times, shape and produce action. Thus there seems to be a non-contingent and non-fortuitous relationship between deliberation on normative reasons for action and action.[19]

The two perspectives on explanation may potentially conflict, as they seem to diverge in the attitudes of agents taken to explain action. From the intentional perspective it is the desire and the means-end belief of the agent that do the explanatory job. From the deliberative perspective, however, it is the attitude towards the normative

---

[17] Smith in [87, pp. 92–129].

[18] Smith in [87, pp. 131–133] for an outline explanation from the intentional perspective.

[19] Smith in [87, pp. 131–133] for an outline explanation from the deliberative perspective.

reason and the means-end belief that explains action. Assuming that accepting a normative reason to p involves an attitude of valuing something, deliberative explanation may be said to trade on the attitudes of valuing and believing. The crucial question in order to clarify whether the two perspectives are in conflict now presents itself: is valuing a way of desiring?[20]

Valuing as a desire or a belief

The attitude of valuing cannot simply be identified with that of desiring.[21] Thus valuing and desiring may come apart. One may desire what one does not value and value what one does not desire. Examples of the former are drug addicts struggling to withstand their desires or defeated squash-players desiring to hit their opponent.[22] Examples of the latter could be found among people suffering from *accidie*, inability to concentrate, physical infirmity, general apathy etc. They may thus lack a desire to do what they value.[23]

These counter-examples also apply to the identification of valuing with a particular mode of desiring, such as desiring something on the basis of reflection and experience.[24] Thus it is not obvious that the drug addict and the furious squash-player are desiring what they would not desire on the basis of reflection and experience.[25] To make it even worse, however, it is conceptually impossible for agents to value what they do not desire. Since this seems to be a clear possibility, this notion of valuing fails and must be abandoned.

In terms of the relationship between motivating and normative reasons, the possibility of a divergence between desiring and valuing has an important consequence, namely that having a motivating reason does not imply the acceptance of a corresponding normative reason, and vice versa. Hence having a normative reason is not relative to an agent's first-order desires, as a Humean theory of normative reasons would have it.[26]

---

[20] Smith in [87, pp. 132–133].

[21] Davidson in [18, p. 86]. Cf. also Smith in [87, p. 138].

[22] Frankfurt in [29, pp. 87–89] and Watson in [99, pp. 100–102]. Cf. Smith in [87, p. 134].

[23] Stocker in [94, pp. 744–745]. Cf. Smith in [87, pp. 120–121 and 135].

[24] Gauthier in [31, pp. 22–23]. Cf. Smith in [87, p. 141].

[25] Cf. Smith in [87, p. 142] for ways around this problem.

[26] Smith in [87, pp. 130 and 165–166].

What, then, of the identification of valuing with any higher-order desires? The attitude of valuing cannot be that of desiring to desire either.[27] The most decisive reason for this is that the identification of valuing with a second order desire is arbitrary. If valuing is a higher-order desire, it may just as well be a third-, fourth- or any higher-order desire.[28]

If valuing cannot be reduced to any kind of desiring *per se*, it remains to be considered whether valuing can be identified with a kind of believing. However, the attitude of valuing cannot be reduced to that of simply believing something to be valuable, since the attitude of valuing seems to imply a favourable attitude towards whatever is valued. Believing something to be valuable is, however, compatible with being indifferent or even being opposed to whatever is believed to be valuable.[29] This argument seems to re-establish a connection between valuing and desiring. After all, desiring clearly involves a favourable attitude to the object desired.

It seems as though we are left with the conclusion that the attitude of valuing can be neither a desire nor a belief without reference to a desire. It may, however, be a belief as long as it expresses something in terms of an agent's desires. In fact, if the attitude of valuing is to be the equivalent of entertaining a belief, then the belief must express something in terms of an agent's desires, since the attitude of valuing involves a favourable attitude similar to the attitude of desiring something. Thus the task is to find a concept of valuing that relates believing and desiring in the manner sketched out here.

Valuing as a belief regarding desires

One way to accommodate this would be to claim that the attitude of valuing is the equivalent of believing something 'desirable to desire'.[30] This cannot be the case, it seems. Thus what it is desirable to do may diverge from what it is 'desirable to desire' such that it is neither desirable to desire what it is desirable to do nor desirable to do what it is 'desirable to desire'.[31] If, for example, one holds that it is desirable to promote one's own long-term self-interest, then it is not necessarily desirable that one

---

[27] Frankfurt in [29, pp. 87–88]. Cf. Smith in [87, p. 142] for the critique of this position.

[28] Lewis in Smith [87, p. 146]. The identification of valuing with the highest-order desire is ruled out because that would rule out the possibility of coming to desire to value differently.

[29] Lewis in Smith [87, pp. 147–148].

[30] Johnston in Smith [87, p. 149].

[31] Smith in [87, pp. 149–150].

desires to promote these interests.[32] These interests may be better served by solely desiring to promote the long-term interests of other people – without this desire being a derivative of a overriding desire to promote one's own long-term self-interests. But if this is granted, then it is clearly not the case that what it is 'desirable to do' (i.e. promote one's own long-term self-interest) is also 'desirable to desire' (i.e. promote the long-term interests of other people), nor is it 'desirable to do' what it is desirable that one desires to do (i.e. promote the long-term interests of other people), since that would be to act against what is found to be 'desirable to do' (i.e. promote one's own long-term self-interest). Thus the example shows that what we value may diverge from what we believe to be 'desirable to desire' – hence an adequate concept of valuing is still required.

According to the position taken on the moral problem we are presenting here, the attitude of valuing is a way of expressing what an agent would desire in certain circumstances. More specifically, it takes an agent valuing something to be the equivalent of an agent believing what she would desire to do in certain circumstances if fully rational, where an agent is fully rational if, and only if, she has no false beliefs, has all relevant true beliefs and deliberates correctly.[33]. Hence, to refresh the intimate link between normative reasons and the attitude of valuing,[34] an agent accepting to have a normative reason to p in the circumstances C is the equivalent of an agent believing that she would desire to p in circumstances C if she had no false beliefs and all relevant true beliefs, and if she was to deliberate correctly.

The requirement of full rationality is not surprising. If an agent valuing something is to be the equivalent of an agent holding a belief concerning a hypothetical desire, it seems wrong to allow for this hypothetical desire to originate in a mistake. If an agent valuing something – i.e., an agent finding herself to have a normative reason for action – is to be construed as an agent holding a belief regarding a hypothetical desire, then it is clearly not the case that this agent is believing something in terms of what she would desire if she was possibly mistaken; on the contrary, the agent is clearly believing something in terms of what she would desire if she was not mistaken. At the core of this observation lies the notion of a normative reason for action. Thus the observation is valid precisely because it seems wrong to claim that an agent has a normative reason to p in circumstances C, if the agent's hypothetical desire to p

---

[32] Parfit in [75, pp. 5–7].

[33] Smith in [87, p. 156ff.].

[34] Cf. our consideration in the above paragraph on intentional and deliberative explanations.

in circumstances C was possibly mistaken. It is simply part of our ways of speaking about, and thereby mastering, the notion of a normative reason, that only those count as normative reasons that trace the most wise or rational thing to do in a situation, where wisdom and rationality have to do with practical knowledge of situations and people.[35]

This point is also readily illustrated by means of a few examples. For instance, an agent desiring cake does not have a reason to eat the plastic explosives that are served on the basis of believing them to be a cake. The agent clearly lacks a reason for action precisely because she would not desire the plastic explosives if she had no false beliefs. Or, to give another example, an agent desiring cake does not lack a reason to eat the cake next door, although she lacks knowledge of the fact of there being a cake in the adjacent room. In this case the agent obviously has a reason for action exactly because she would desire the cake next door if she had all true beliefs.[36]

The overall conclusion to be drawn on the basis of the considerations in this and the previous paragraph is thus that an agent valuing something – i.e., considering herself to have a normative reason to act in a certain way – may be considered to be the equivalent of an agent believing she would have a desire to act in a certain way in the relevant circumstances if she was fully rational. This partly reflects the fact that in our use and mastering of language there is notion of a normative reason for action according to which an agent has a normative reason to p in circumstances C if and only if she would desire to p in circumstances C if she was fully rational.[37] Note that this notion of a normative reason contradicts a Humean concept of normative reasons, according to which one has a reason for action if the action would satisfy a desire.[38] As such, this account of normative reasons is clearly in opposition to a Humean approach to the explanation of behaviour based on deliberation.

---

[35] Smith in [87, pp. 150–152]. Cf. also [87, pp. 29–32 and 39–41] for the notion of platitudes as being involved in coming to master linguistic terms.

[36] Smith in [87, pp. 157–162] and Williams in [101, pp. 101–105].

[37] Smith in [87, pp. 150 and 156]. Note, very importantly, that it seems conceptually possible that a fully rational agent may not desire anything in a given situation while at the same time having reasons for action. Adjustments may be needed.

[38] Smith in [87, p. 130].

Resolving the moral problem

Assuming the concept of valuing developed in the previous paragraph, it follows that if an agent finds herself to have normative reason to p, then the agent believes she would desire to p if she was fully rational. Imagine now an agent believing she would desire to do p if she was fully rational, but failing to have the desire to p. Given that the mere fact of actually desiring not to p does not provide any reasons for changing an evaluative belief, she fails to have a desire which she believes it is rational for her to have. Thus she is irrational by her own standards. The crucial lesson is that if an agent believes she would desire to p if she was fully rational, then she rationally should desire to p. That is, rationally the agent should endeavour to acquire a desire to p.[39] Hence we get: If an agent finds herself to have a normative reason to p in circumstances C, then she rationally should desire to p in circumstances C.[40] In other words, holding a belief in a normative reason to p should, normatively speaking, lead to (cause) a desire to p. This is not only a conceptually possible but actually a normatively required, transition.[41]

The grounds needed for outlining the second position on the moral problem have now been provided. We started this subsection by saying that deliberative explanations are genuine explanations of intentional action – in short, that deliberations may motivate intentional action. We have subsequently seen that valuing – the explanatory attitude in deliberative explanations – is a kind of believing which may produce a desire to act in a certain way in the rational agent. Assuming that a belief and a desire are sufficient to produce action, we thus conclude that an agent's beliefs regarding her normative reasons for action may motivate action. Moreover, if an agent believes herself to have a moral reason to p, then this belief may motivate action in the absence of practical irrationality.

The second position on the moral problem has three components. First, it tacitly maintains cognitivism. Second, it defends a version of appraiser internalism by claiming that if an agent believes herself to have a normative reason, then this belief may motivate action, but it denies that this belief necessarily motivates, as an agent may suffer from practical irrationality. Third and finally, it denies that the Humean theory of motivation applies to normative reasons. Although it allows for

---

[39] Smith in [87, pp. 177–178].

[40] Smith in [87, pp. 148, 178 and 181].

[41] Smith in [87, p. 179].

an independent existing desire to be a prerequisite of action – and thus confirms the Humean claim that all actions are produced by desires – this desire is itself produced by a belief. In other words, the motivation originates solely in the belief of having a normative reason.[42]

This solution to the moral problem has clear-cut implications for the explanation of intentional action. Intentional action may now be explained by deliberation in itself. If an agent through a process of deliberation comes to believe that she has a normative reason to p, then this belief may cause her to have a desire to p and thus in turn cause her p-ing. Note, first, that this explanation is done in normative terms. An agent's belief may produce a corresponding desire because it is rational for the agent to have the desire, given her belief. Note, second, that if intentional action may now be explained solely by reference to the moral deliberations of an agent, then the explanation of intentional action on the basis of a Humean theory of motivation is limited to those cases in which an agent does not act on the basis of deliberations. Thus the solution advocated implies that some actions may still be motivated by the combination of a belief and a desire in the Humean fashion – i.e., the Humean theory of motivation still holds for motivating reasons.

3.1.4 Pure cognitivist internalism

A third approach to the moral problem solves it by simply writing off the Humean theory of motivation in favour of a pure cognitive theory of motivation: i.e., cognitivism and some form of appraiser internalism.

Two cognitive representations

The position in question is pure in the sense that it claims there is only one distinct kind of motivation. More specifically, it holds that not just moral action but any kind of purposive action is always motivated by two basic cognitive representations of the world taken to be the only necessary states for action to occur.[43] These representations consist of a representation of how the world is at present and a representation of how the world will be if and when a relevant action is successfully completed, where the latter is to be taken as a subjective conditional of the form 'If I were to act in such

---

[42] Smith in [87, p. 179]. Dancy in [16, p. 9] labels theories in which the desire is somehow produced by, for instance, a belief 'motivated desire theories'.

[43] Dancy in [16, pp. 18 and 29].

and such a way, this would be the result'.[44] Hence the first representation specifies what an agent is working from – the circumstances necessitating action. The second specifies what the agent is working to achieve.

Since these representational states are the only states necessary for an agent to be motivated, it remains to be clarified whether or not they have the character of beliefs or desires or both. For the purpose of deciding this, a distinction between the direction of fit of cognitive states is needed. The 'direction of fit' of cognitive states refers to the relation between the content of the cognitive state and the world. A belief, for instance, is a state that aims to be caused by the truth of its own content, whereas a desire is a state that aims to cause its own content to be true.[45] Armed with this characterization of cognitive states such as belief and desire, we can now examine the direction of fit of the representational states motivating action in order to decide whether they are comparable to beliefs or desires.

The direction of fit of the first of the representations seems quite uncontroversially to be that of a belief. The direction of fit of the second representation presents more problems. Thus the Humean would clearly claim that an agent being motivated in part by the presence of the second representation must be acting so as to make the content of the second representation – i.e., the subjunctive conditional – true. The obvious consequence is that the second representation would be a state aimed at causing its own content to be true, and hence have the direction of fit of a desire. However, this way of interpreting an agent's action based on the second representation is not the only one available.[46] One could reasonably claim that an agent being motivated in part by the presence of the second representation has the intention of making the consequent of the subjunctive conditional true. Hence the second representation cannot be said to be a state that aims to cause its own content to be true – only part of it. This does not, however, rule out the possibility of the second representation having the direction of fit of a desire. Thus it may be argued that if an agent being motivated in part by the presence of the second representation is somehow the equivalent of an agent being motivated on the basis of an intention to make the consequent of the

---

[44] Dancy in [16, pp. 14 and 28–29].

[45] Dancy in [16, pp. 27–29]. A correspondence theory of truth is tacitly assumed.

[46] Dancy in [16, p. 29] notes that there are more ways of acting consistent with acting so as to make the subjunctive conditional true. Dancy seems to hold that this somehow negates the possibility of acting to make the subjunctive conditional true, possibly because a certain action then would be underdetermined by its motivating states.

subjunctive conditional true, then the second representation has the character of a desire, as the intention will be a state with a content that motivates an agent towards making the content true.

The answer favoured by the theory here presented is to distinguish between motivation and intention. The two representations are what motivate an agent. The intention does not motivate – it is simply the agent being motivated by the two representations. In being motivated by the two representations an agent simply acts with a certain intention (to make the consequent of the subjunctive conditional true).[47] Having paved the way, the theory here presented concludes that the two cognitive representations motivating action have the character of beliefs in virtue of having the direction of fit of beliefs. 'In fact, nothing is lost by thinking of them as beliefs.'[48] It now remains to be settled how a concept of desire fits into this framework.

The role of desires in the motivational process

Several conceptions of desires and their role in the motivational process are compatible with the concept of motivation developed above. A desire may be something that is simply ascribed to an agent acting on the basis of the two beliefs, or it may be either of the two representations or it may be conceived to be the state of being motivated by the gap between the two representations.[49]

In the first case, to have a desire to p is simply to be motivated by the representation to p (or to q if the agent believes that he p's by q-ing): i.e., the desire is consequentially ascribed.[50] The second holds two possibilities, of which the identification of desire with the second of the two representations seems the most natural. Either solution would be awkward since it would entail that the relevant representation should have both directions of fit: as both a desire and a belief. The third case, in which a desire is taken as the state of the agent being motivated by the gap between the two representations, is seemingly the most promising, as it allows for the desire to

---

[47] Dancy in [16, pp. 19 and 29].

[48] Dancy in [16, p. 29].

[49] Dancy in [16, pp. 19–20] for the possible roles of the concept of desire within this theory. The short elaboration to follow is based on the same pages.

[50] Dancy in [16, p. 9]. Dancy labels this kind of theory 'pure ascription theory'. They are contrasted with 'motivated desire theories' (see note 42 on page 45).

be an independently existing, non-cognitive state without it being causally effective in the production of action.[51]

Although there are several conceptions of desire available for this theory of moral motivation, it remains to be shown how the theory can account for phenomena such as weakness of will. Weakness of will may be seen in a case of two agents sharing a belief in the morally right action but only one of them being motivated to act on the basis of this belief.[52] Obviously, one way of accounting for this phenomenon would be to resort to a Humean theory of motivation and claim that the difference between the two agents is a difference in desire. For a pure cognitive theory of motivation this is evidently not a feasible manoeuvre. And thus weakness of will comes to pose a challenge.

Intrinsically motivating states

The solution here favoured is founded in the denial of the claim that, if a state is sufficient motivation in circumstances C_0, it will be sufficient motivation in any circumstances C_1, C_2, C_3 ... C_n.[53]

The corresponding positive claim is developed by means of two important distinctions. The first is a distinction between states that necessarily motivate – in other words, states that cannot be present without motivating – and states that only contingently do so – in other words, states that can be present without motivating. The second is a distinction between states that motivate in their own right and states that do not motivate in their own right. Reverting to the exposition of the positive claim, this may now be stated as the assertion that there exist states that motivate in their own right, although they do not necessarily do so. These states are dubbed intrinsically motivating states.[54] Given that moral beliefs or judgements regarding the morally right action are made up of two cognitive representations which are themselves an intrinsically motivating state, it clearly follows that a pure cognitive theory of motivation can concede the possibility of weakness of will while at the same time maintaining a form of internalism. Moral beliefs only contingently motivate, but they do it in their own right. For this salvaging move to be acceptable, however, the coherence of the concept of an intrinsically motivating state must be established. In trying

[51] See also Dancy in [16, p. 29].

[52] Dancy in [16, p. 22].

[53] Dancy in [16, p. 22].

[54] Dancy in [16, p. 24].

to do this, we may also get an idea of the difference between the agent suffering weakness of will and the agent not so suffering.

The coherence of the concept of an intrinsically motivating state rests on the coherence of the claim that what counts as a motivating reason in one case may not count as such in another case, owing to changes in background conditions that are not themselves part of the motivating reasons but which may be part of the causes of the action.[55] This claim is clearly opposed to a Humean theory of motivation, according to which the motivating reasons are sufficient for the production of action – in other words, the motivating reason for action and the cause of action are simply the same.

Let us take an example. Imagine someone who runs a marathon everyday because she believes it builds character and good health. She further believes that running a marathon a day does not affect her resources available for other kinds of work. In this case it seems fully reasonable to interpret the agent's motivating reason as being simply and solely the belief that running a marathon a day builds character and good health. Even though the agent might loose her motivation if she was to believe that the running decreases the general resources, it seems mistaken to count the belief that the general resources are not so affected by running among the motivating reasons for action. The action is intelligible without this belief, and furthermore the belief is in itself an odd reason for running a marathon a day.[56] On the basis of this example it seems as though the background conditions, such as beliefs, may influence a given reason's ability to motivate without being part of the motivating reasons. That is, the background conditions are part of the conditions sufficient for the production of an action – i.e., the cause of an action – without being part of the motivating reasons for action. The concept of intrinsically motivating states thus seems coherent.

Motivating beliefs

So far several important properties of the cognitive representations in question have been established – for instance, that they have the direction of fit of beliefs and may be conceived of as intrinsically motivating states. It still remains to be shown, however, that beliefs may motivate at all. After all, it seems as though beliefs, given their direction of fit, are not sufficient to motivate action simply because they do not convey reasons for changing anything. The problem is this: how can beliefs motivate

---

[55] Dancy in [16, p. 24].

[56] Dancy in [16, p. 25].

any actions if they do not convey any reasons for action? To be motivated to act in a certain way, one must have a reason to act in this way – one must have a reason that beliefs seem unable to convey. This claim is not as innocent as it appears. If a belief, a cognitive state with a content representing the state of the world, is not to provide any motivation for action because of not conveying any reasons for action, then it seems to presuppose that the world does not in itself make a difference to how one should act. It simply presupposes that the world is inert in terms of providing reasons for action.[57]

The inertness of the world may be grounded in Cartesian metaphysics, according to which the significance and importance of an event in the world are taken to be the contribution of a mind – either of our own or of some other being (e.g., God). Put slightly differently, when we find that an event carries a certain significance and importance and is therefore a reason for action, this is, in the eyes of the Cartesian, either the gift of our own mind or the recognition of the bestowal of significance and importance on it by another mind.[58] The world does not present itself as a reason for action: the reason for action is something that is added by a mind in its interaction with the world. Through rational, intellectual contemplation of the features of the world and our position in it we may come to discover reasons for action. It follows from this picture that there can only be two directions of fit of cognitive states, and, moreover, that no state can have both directions of fit.

Given this outline of how a Cartesian metaphysics underlies the adduced problem, it seems obvious that the pure theory is committed to a non-Cartesian metaphysics.[59] More specifically, the pure theory holds that facts may convey reasons for action, and thus, if believed, become a source of motivation. Consequently, it is not the believing that is in itself the source of motivation. Rather, it is the fact believed that is the source of motivation or, to put it slightly differently, that initiates motivation. Motivation thus comes to have its source in, or to be initiated by, facts and not in beliefs: i.e., intrinsically motivating beliefs thus have their origin in facts intrinsically providing reasons for action.

To conclude, the pure theory, like the Humean theory of motivation, ultimately implies that beliefs are motivationally inert[60] – motivation requires beliefs, but the

---

[57] Dancy in [16, pp. 31–33].

[58] Dancy in [16, p. 31].

[59] Dancy in [16, p. 32].

[60] Dancy in [16, p. 32].

source of motivation is not the beliefs. Contrary to the Humean theory of motivation, it relegates the source of motivation not to a Humean desire but rather to facts of the world. In terms of direction of fit the two representations maintain the direction of fit of beliefs, as they are still states that aim to be caused by the truth of their own content. However, they have both directions of fit in the significant sense that they are both representations of the world and also reasons for changing the world.[61] So far, so good.

3.2   Explaining the basic premise

Having considered some different conceptions of moral motivation developed in response to the moral problem, it is now time to clarify how these considerations may feed into a model for the explanation of the basic premise ([TBP]). Before sketching such a model, we will, however, make a basic investigation into the role of beliefs and ontological conditions in the explanation of human intentional behaviour. The explanatory model developed on the basis of this investigation is logically independent of the different conceptions of moral motivation already considered. This is a significant result since it means that we can pursue an explanation of the basic premise without these endeavours being open to objections aimed at some underlying conception of moral motivation.

3.2.1   The role of beliefs in explanation

For the purpose of clarifying the role of beliefs in the explanation of human intentional behaviour we will make a distinction between different kinds of belief and then investigate how these beliefs may enter into an explanation of human behaviour according to each of the conceptions of moral motivation already considered.

Three kinds of beliefs

It seems as though any agent's deliberation as to how to act in a given situation involves at least three kinds of belief. First of all, it involves beliefs concerning the properties of the situation in which the agent is to act – that is, beliefs concerning the set of properties possessed by the entities constituting the situation, and thus beliefs

---

[61] Dancy in [16, p. 33–34].

regarding the kind or type of situation. We could posit, for example, the agent's belief
that there is a starving child in front of her, and the belief that she is in possession of
food. Second, it involves beliefs of a general character relevant to the deliberation of
how to act in the situation in that they relate to an entity believed to constitute the
situation. This might, for example, be the belief that food may remove hunger. Note
that these beliefs are, so to speak, brought to the situation and that their relevance
is established on the basis of the first kind of belief, namely on the basis of beliefs
concerning the set of properties of situations. Third, it involves a belief concerning
what the agent is normatively required to do. These are beliefs specifying what the
agent has a normative reason to do in the relevant circumstances or situation, where
this reason is generated by a normative system. This might be, for example, the belief
that she ought to give the food she is in possession of to the starving child in front
of her.

In short, an agent deliberating how to act in a given situation may reason on the
basis of, on the one hand, background beliefs establishing the kind of situation and
background beliefs of relevance for situations of this kind, and, on the other hand,
beliefs concerning her normative reasons for action in situations of the relevant kind.

At this point it is interesting to revisit the different approaches to moral motivation
examined in the previous section. There we described how the Humean approach takes
a desire to play the main role in the production of intentional action and how the
non-Humean approaches converge on the view that a belief in a normative reason for
action may play the main role in the production of intentional action. In the light of
the distinction between kinds of beliefs just made, it might be asked what role, if any,
background beliefs play in an explanation of human intentional behaviour according
to these approaches.

Background beliefs and means-end beliefs

The Humean approach incorporates fairly straightforwardly background beliefs as
defined above in the conception of moral motivation. The Humean construes moral
motivation as consisting of an appropriate desire and certain means-end beliefs, where
desire and beliefs are distinct existences. The agent being motivated to do q in circumstances C is consequently seen as desiring to p and believing that he would p by
doing q in circumstances C. The agent desiring to help the poor and needy may consequently come to be motivated to give his food to the starving child in front of him
if he holds the means-end belief that he would help the poor and needy by giving his

food to the child in front of him. As should be clear, a number of background beliefs underlie the agent's means-end belief in this. Thus the agent's belief that he may help the poor and needy by giving his food to the starving child in front of him clearly rests not only on beliefs concerning the character of situation he is in – namely, that there is a starving child in front of him, that he is in possession of food and so on – but also on more general beliefs, such as the belief that food may remove hunger.

This dependency between an agent's means-end beliefs and his background beliefs clearly shows a dependency between an agent's motivation to do q in circumstances C and the agent's background beliefs. Thus the agent may simply fail to be motivated to do q in circumstances C if he changes any of his background beliefs on which his means-end beliefs are based. In terms of explaining human intentional behaviour it clearly follows that a Humean conception of moral motivation would leave room for the inclusion of background beliefs in such an explanation.

Background beliefs and justification

The first of the non-Humean approaches claims that an agent believing he has a normative reason to p in circumstances C is expressing the belief that he would desire to p in circumstances C if he was fully rational, in other words that he would desire to p in circumstances C if he had all true beliefs, no false beliefs and was deliberating correctly. It is this belief that – in so far as the agent is rational – is taken to motivate action.

The agent considering whether or not to do p in circumstances C must decide what his more rational self would do in the relevant circumstances. That is, he must decide whether or not he would desire to do p in circumstances C if he was fully rational.[62] It seems as though this decision involves – either directly or indirectly – the forming of two kinds of second-order beliefs: on the one hand, a belief regarding what he would believe in the circumstances C if he was fully rational (i.e., a belief as to which beliefs would be in the set of beliefs made up of all true and no false beliefs) and, on the other hand, a belief regarding what he would desire to do in circumstances C if he was fully rational. In forming both of these second-order beliefs an agent is entering into a process of justification. Thus to decide what the agent would believe and desire in circumstances C if he hypothetically had all and only true beliefs must involve the agent giving reasons why he in the relevant circumstances would believe

---

[62] Smith in [87, pp. 153–154].

that p rather than $\neg p$, and desire to p rather than $\neg p$, where this justification is taken to be tracking the truth of the matter. In short, to decide what he would desire in circumstances C if he had all and only true beliefs must involve giving reasons to the effect that his beliefs regarding the circumstances C are true or false as well as giving reasons to the effect that his actual desire in the circumstances C is justified or not so justified.[63] It is at exactly this point background beliefs seem to become relevant. Thus background beliefs may provide the material for the justificatory process as both the beliefs requiring and providing justification.

Imagine an agent trying to decide what he would desire in circumstances C if he was fully rational. As in the example above, the agent holds the beliefs that there is a starving child in front of him, and the belief that he is in possession of food. He also holds the general belief that food removes hunger. Furthermore, his present desire is to eat the food himself. For the agent to decide what he would desire in circumstances C if he had all and only true beliefs must initially involve giving reasons in support of his relevant beliefs: first, in support of there being a starving child in front of him, for example, that the child in front of him shows sign of being feeble and undernourished and so on; second, in support of him being in possession of food, for example that he is carrying a lunch box etc.; third, in support of the general beliefs of food being able to remove hunger, such as that food may be broken down into certain components that the human body needs. Given that the agent on the basis of these beliefs desires to share his food with the starving child, the next step is then to test whether this desire would be held under conditions of full rationality: in other words, to try to justify the belief that this desire would be held under conditions of full rationality. This belief may, for example, be supported by the belief that the agent generally desires to help other people. However, the end-result of this process may very well be the justification of a belief to the effect that a desire different from the present desire would be held under conditions of full rationality.[64]

The lessons to be learned from these reflections are the following. First, an agent's deliberation on his normative reason for action in the conceptual scheme of the first

---

[63] Smith in [87, pp. 164–174.] Smith defends a non-relative conception of justifying desires according to which the justification *per se* is not relative to an agent. That is, it may be discussed whether p supports q in the sense that the proposition p may or may not count in favour of q, but not in the sense in which it is discussed what the relation of support in itself amounts to. Cf. [87, p. 168].

[64] Cf. also Smith in [87, p. 155].

of the non-Humean approaches involves the agent's background beliefs. In order to decide what he has a normative reason to do in circumstances C, the agent must form and justify beliefs regarding the properties of the relevant situation, apply justified general beliefs with relevance for the situation, and form and justify beliefs regarding his desire under conditions of full rationality. Since justification involves an agent's background beliefs, the decision-making process thus comes to include background beliefs as both beliefs to be justified and beliefs that justify. Second, the motivational power of an agent's beliefs in having a particular normative reason for action is not dependent on the agent's background beliefs, in the sense that changes in these affect the motivational power of the belief in the particular normative reason. However, given a decision-making process along the lines just outlined, it seems clear that changes in background beliefs may lead to changes in what an agent believes he has a normative reason to do.

Again, if background beliefs may in this way influence what an agent is motivated to do, it clearly follows that an agent's background beliefs are relevant for an explanation of an instance of action based on a process of deliberation.

Background beliefs and representations of the world

The second non-Humean approach holds that the motivational power of an agent's belief in having a reason to p is directly conditional on background beliefs, in the sense that alterations in these beliefs may directly affect the ability of an agent's belief in having a reason to p to motivate – without, however, these background beliefs being counted among the agent's reason for action. In short, a belief in a reason for action is an intrinsically motivating state.[65]

The direct conditionality between background beliefs and motivation seems to be grounded in the conception of motivation. As has already been seen, the second non-Humean approach construes moral motivation as originating in two cognitive representations The first is a representation of how the world is at present and the second a representation of how the world would be if the agent were to act in a given way, where both of these representations have the direction of fit of beliefs. It is these representations that form the intrinsically motivating states – that is to

---

[65] Dancy in [16, pp. 24 and 25]. Bear in mind that the second of the non-Humean approaches does not distinguish between normative and motivating reasons for action. Hence all references will only be to reasons.

say, the states that motivate in their own right but without motivating necessarily.[66] From this account of the origins of moral motivation it is evident that the second of the non-Humean approaches ties together an agent's moral motivation and her background beliefs. Thus the representation of the world as it is simply amounts to a representation of the properties of the circumstances, where this representation has the direction of fit of beliefs. Likewise the second representation of how a given action would influence the world trades on representations of the workings of the world, where these representations also have the direction of fit of beliefs. The obvious implication of this interpretation is that an agent's motivation becomes directly vulnerable to changes in background beliefs. Changes in one set of beliefs may obviously impact on the power of other beliefs to influence our behaviour. Thus changes in background beliefs may affect the power of the two cognitive representations to motivate action.

Reverting to the example above, let us imagine an agent trying to decide what to do in circumstances C. In terms of the first representation, the agent is representing the world as containing both a starving child positioned right in front of her but also food in her possession. Furthermore the agent represents the world as encompassing a mechanism according to which food may remove hunger. In terms of the second representation, the agent represents the outcome of sharing her food with the child in the particular circumstances as a state of the world in which the child would suffer less from starvation. The second non-Humean approach claims that an agent embodying a set of representations of this kind may (or will) be motivated to act accordingly, given certain background conditions.

Although the set of representations used in the example above may have to be extended to be adequate, the example quite clearly illustrates how an agent's background beliefs are linked to her motivation. It is evident that if the agent in the example changes some of her beliefs regarding the properties of the situation or circumstances – for example, that the child is not really starving – then this directly affects her motivation. Likewise, if the agent comes to hold additional beliefs – for example, that relieving the child of its hunger would be at the unwanted expense of the agent's own children – then this would also directly affect her motivation. Thus changes in relevant beliefs directly influence motivation because of the implied change in the conditions sufficient for action to occur – i.e., because of the implied change in the cause of action.

---

[66] Dancy in [16, pp. 23–24].

Having shown how the two cognitive representations of the world are vulnerable to changes in background beliefs, we need somehow to clarify how this relates to an agent believing herself to have a reason for action. In the outline of the second of the non-Humean approaches at the beginning of this subsection we made a distinction between the causes of an agent's action and the agent's reason for action. An agent may believe herself to have a reason for action without this belief being able to motivate action. For a belief in a reason for action to motivate action requires appropriate background conditions, which include the agent's background beliefs. Thus the cause of action is both an agent's belief in a reason for action and a set of background conditions including a set of background beliefs.

Let us briefly consider an interesting outcome of our investigations. The noteworthy feature is the implied role of deliberation in the process of coming to act on the basis of a belief in a reason for action. The second of the non-Humean approaches seems to allow for deliberation to play a role linked to the role of background beliefs. Thus it seems as though deliberation may serve the function of an intellectual filtering of an agent's background beliefs, a filtering that decides whether the agent holds any beliefs of relevance for the situation in question. Since the ability of a belief in a reason for action may be undercut by background beliefs, deliberation may clearly serve the role both of clarifying if there are any such beliefs in an agent's web of beliefs and of justifying these beliefs in the face of a believed reason to the contrary. In the case of an agent believing she has a reason for action the motivational power of which is undercut by another belief, the agent must decide whether she is misinterpreting her reason for action or her other belief is false. In the second of the non-Humean approaches, therefore, deliberation may also come to serve a justificatory purpose and in this way come to affect an agent's motivation. This is not, however, in the sense of the first of the non-Humean approaches, in which deliberation is the origin of an agent's belief in a normative reason for action. (There is further discussion of this point below.)

Once again, we may conclude that if background beliefs may influence what an agent is motivated to do, it clearly follows that an agent's background beliefs are relevant for an explanation of an instance of action.

Beliefs and the non-Humean approaches

Before completing these considerations on the role of background beliefs for the explanation of human intentional behaviour, let us briefly elaborate on the difference in the role of background beliefs in the two non-Humean approaches.

A first noteworthy difference between the first and second of the non-Humean approaches has to do with the way in which they tie together background beliefs, motivation and the belief in having a reason for action. In the first of the non-Humean approaches changes in relevant background beliefs may affect the justification for a belief in a normative reason for action and may consequently lead to a change in what the agent believes she has normative reason to do and hence, in the end, to a difference in what the agent is motivated to do. It is the agent's belief in a normative reason for action, however, that carries the motivational power – i.e., an agent is motivated to act differently only insofar as she changes her belief as to what she has a normative reason to do. In the second of the non-Humean approaches an agent can, it seems, maintain a belief in having the same reason for action in two situations in which her motivation differs because of a difference in her background beliefs.

A second noteworthy difference between the two non-Humean approaches relates to their account of an agent's formation of the belief in a reason for action. Although we have explicitly focused on the role of background beliefs for motivation in the preceding paragraphs, they have also partly revealed a difference in non-Humean accounts of how an agent may arrive at belief in a (normative) reason for action. In the first of the non-Humean approaches an agent's belief in a normative reason for action is the result of careful deliberation of what the agent would desire in the relevant circumstances if she were fully rational. As such, the belief in a normative reason for action is the outcome of an intellectual exercise drawing heavily on the agent's beliefs regarding the relevant circumstances, human nature and so on. The second non-Humean approach, by contrast, does not depict an agent's belief in having a reason for action as the outcome of deliberation. However, what an agent is motivated to do may be the result of deliberation in the sense that an agent may wonder whether she has any beliefs that in the present circumstances count against acting on the basis of what she believes she has a reason to do. This difference relates directly to the non-Cartesian metaphysics of this approach. As outlined earlier, the second non-Humean approach holds that the world may present itself to us as reasons for action, and thus in appropriate circumstances may come to be the cause of action. That is, the world may simply come to act in and through an agent. (For further discussion of this point, see below.)

Although the role of the background beliefs is different within the two approaches, they both maintain that the background beliefs may influence what an agent finds herself to be motivated to do. Thus in order to explain an instance of intentional behaviour both approaches imply that an agent's background beliefs must be cited.

3.2.2   The role of ontological conditions in explanation

Not only background beliefs but also the ontological background conditions may enter
into an explanation of human behaviour. As a first observation one may distinguish
between those ontological background conditions that are psychological and those
that are non-psychological in nature. In what follows we will start with the latter
by considering the relationship between an agent's belief and the world, and more
specifically the relationship between the belief in a reason for action and the world.
Subsequently the former will be investigated by considering how the agent's psychology may also influence his or her motivation. All of these investigations are, of course,
carried out in the context of both the Humean and the non-Humean approaches to
moral motivation.

Ontological conditions and background beliefs

As indicated, the role of the non-psychological circumstances in which an agent
acts seems to hinge on the role of an agent's beliefs. Thus it seems as though non-
psychological circumstances may be primarily relevant for the explanation of human
intentional behaviour by virtue of an alleged ability to influence what an agent comes
to believe. Disregarding undue philosophical scepticism, it seems a plausible claim
that the content of many, but not all, of our beliefs originates in and is predominantly
shaped by some sort of an external world impinging on us. From this seemingly external world we receive sensory stimuli, which by means of complex machinery lead us
to hold certain beliefs regarding the furniture of the world and our position in it. This
rudimentary picture of the relationship between the external world and our beliefs
regarding it has important implications.

In terms of explaining human intentional behaviour it follows that the explanation
of an instance of human behaviour may ultimately refer to non-psychological circumstances as the possible origin of the background beliefs that may directly or indirectly
influence what an agent believes he has a normative reason to do. Thus in the Humean
approach the non-psychological circumstances would enter into the explanation of an
instance of deliberative behaviour as the possible origin of those background beliefs
tied to the agent's means-end beliefs. In the first of the non-Humean approaches
the non-psychological circumstances would enter into an explanation as the possible
origin of the background beliefs involved in the agent's beliefs concerning what he
would believe and desire if he was fully rational. In the second of the non-Humean

approaches the non-psychological circumstances would enter into an explanation as the possible origin of the cognitive representation of how the world is at present.

It follows, not surprisingly, that in both the Humean and the non-Humean approaches an agent's belief in having a normative reason for action is related to the properties of a situation in such a way that, were the properties of the situation to be different, then the agent might fail to be motivated to act in a certain way – either because of changing his means-end beliefs, because of changing his beliefs concerning what he has normative reason to do or simply because of the motivational power of his belief in a normative reason for action being undermined.

It is worth noting that the second of the non-Humean approaches further develops the relationship between an agent's belief in having a normative reason for action and the world. By means of discarding a Cartesian metaphysics in favour of a metaphysics that allows for facts or 'the world' to convey reasons for action, the second non-Humean conception allows for ontological conditions to play a crucial role in the motivation of action. Thus it may be circumstances themselves that play the main role in motivating the agent's actions. This further supports the conclusion already envisaged: namely, that a model for the explanation of human intentional behaviour must have reference to the properties of the situation in which the agent is acting.

Ontological conditions and moral depression

With regard to the psychology of an agent, both the Humean and the non-Humean approaches acknowledge its role in the explanation of human intentional behaviour. They differ, however, in the way in which they take the psychological circumstances to be relevant for the behaviour of an agent.

As should be clear by now, the Humean conception of moral motivation takes the prime mover of action to be a psychological state, namely a desire. An agent is simply motivated to act in a certain way insofar as the agent has a desire for something linked to the action by means-end relations. Consequently, any model for the explanation of human intentional behaviour must include a reference to the psychology of an agent as a prerequisite of action.

Non-Humean approaches take the psychology of agents into account by allowing for forms of so-called moral depression, such as weakness of will or *accidie*, to undercut an agent's motivation to act on his belief in having a normative reason for action. In the first non-Humean approach this is accounted for by the possibility of an agent suffering from practical irrationality. In the second non-Humean approach it is accounted for by the possibility of the background conditions, together with a belief in a reason for action not being sufficient to motivate action.

The possibility of an agent coming to suffer from moral depression has to be addressed in a model for the explanation of human behaviour. Thus any explanation of an agent's action must address the question of whether or not the explanation is to be given on the assumption of the agent suffering or not suffering from moral depression. That is, any explanation will have to state explicitly whether or not the explanation of an instance of action is carried out on the assumption either that the agent is or that he is not suffering from moral depression.

3.2.3   Explanatory model

In this chapter we have so far tried to account for some of the important views and positions taken in the discussion of how to account for human intentional behaviour. However, as has been made clear in the previous subsection, the different conceptions of moral motivation seem to leave room for background beliefs and circumstances of both a psychological and a non-psychological nature to play a role in the behaviour of an agent in such a way that a change in either of these may come to make an agent act differently. In the conclusion to this chapter we will now pull these threads together in a tentative model for the explanation of human intentional behaviour: a model to be used in the attempt to explain the difference in interaction claimed in the basic premise, [TBP], by reference to a belief in the reality of a particular interacting agent.

A model of explanation

The simple and tentative model to be used in the explanation of why an agent intentionally acted in a certain way is this:

[**Exp**]   (a)   Agent A holding the set of beliefs B was in a situation S
                  with the set of properties P

             (b)   Because of certain properties $p_1, p_2 \ldots p_n \subseteq P$ of the situation S,
                  agent A believed she was in a situation of type C

             (c)   Deliberations on beliefs $b_1, b_2 \ldots b_n \subseteq B$ relevant for situations
                  of type C, led the agent A to be motivated
                  to do x in situation S

             (d)   Agent A deliberated correctly

The first premise, (a), is fairly straightforward. It simply lays out the setting: An agent with certain beliefs is in a particular situation.

Premise (b) states that the agent in question holds beliefs concerning the type of situation because of certain properties of the situation. This is just to say that the content of the beliefs agent A forms concerning the type of situation are somehow grounded in properties of the situation – they are not pure inventions of the mind resulting from the work of a mad scientist or hypnotist. Consequently, (b) is not to be read as saying that the agent's belief was caused by the properties of the situation in a crude way. The crude reading would, it seems, exclude the possibility of the agent being wrong regarding the type of situation: i.e., it would imply epistemological infallibility. The former reading, however, states only that the belief concerning the type of situation was causally affected by the properties of the situation. It is still possible for the agent in forming a belief regarding the type of situation to misjudge and misinterpret the type of situation. Premise (b) is at the core of explaining the basic premise.

Premise (c) is a bit more complicated. Three things are worth noting. First of all, premise (c) is supposed to allow for three readings in terms of the role of deliberation. In a Humean conception of moral motivation the deliberations referred to concern the means-end relations relevant for an agent's ability to gratify a certain desire through a certain action in the particular situation. In the first of the non-Humean approaches an agent's belief in a normative reason for action is arrived at through a process of deliberation concerning what the agent would desire if fully rational. In the second of the non-Humean approaches an agent may arrive at a belief in having a particular reason for action simply on the basis of the state of the world. However, if this belief is to act as a basis for motivation, the agent must not hold any beliefs inconsistent with it. Consequently, deliberation may here be referring to the process through which the agent decides whether or not she holds any beliefs inconsistent with her belief in having a particular reason for action. Second, it is also worth noting that a non-Humean reading of premise (c) tacitly assumes that the agent coming to be motivated on the basis of a consideration of her beliefs relevant for the situation in question was psychologically disposed to act on her beliefs. That is, it tacitly assumes that an agent did not suffer from weakness of will, accidie, apathy or any other form of moral depression. Third and finally, note that in both a Humean and in the non-Humean readings of premise (c) the beliefs referred to as the 'beliefs $b_1, b_2 \ldots b_n \subseteq B$ relevant for situations of type C' are precisely those we earlier subsumed under the generic label of 'background beliefs'.

Premise (d) assumes that the agent in question is deliberating correctly. Reasoning correctly means that the agent is not sloppy in her reasoning. It involves taking all relevant beliefs into account, understanding the implications of beliefs, weighing conflicting beliefs and so on. It does not, however, mean that all of the agent's beliefs are true or tracking the truth. Thus the agent may be deliberating correctly on the basis of false beliefs.

The explanatory model and the basic premise

It is worth briefly considering what has been achieved by introducing the model above. As should be clear from our previous considerations of reasons for action, the model contains premises that, strictly speaking, are not necessary to explain an instance of intentional action but which nonetheless are of great importance for our attempt in this book to build an explanation of why people act differently inside and outside cyberspace.

More specifically, the model links three features of a situation involving action in two steps. First, it links the properties of a situation with an agent's beliefs concerning the type of situation. Second, it links the agent's beliefs concerning the type of situation and her deliberation and beliefs relevant for the type of situation with the motivation, and thus the action, of the agent. Both of these steps are vital for our attempt to explain the basic assumption. Thus the basic premise states that the difference between interaction inside and outside cyberspace applies to cases in which interacting agents decide their course of action on the basis of deliberation, and in which an agent's behaviour reflects the properties of situations such that a difference between the actions in two situations entails a difference between the properties of the situations. The second step in the explanatory model – the linking of beliefs and deliberation – thus straightforwardly reflects the fact that we are to explain (a difference in) actions resulting from the deliberations of agents. The first step in the explanatory model – the linking of the properties of situations and the agent's beliefs – thus reflects that we are to explain a difference in actions reflecting a difference between situations.

It is also worth noting that the explanatory model takes into account another aspect of the basic premise. Thus the basic assumption actually states that the difference between interaction inside and outside cyberspace applies to cases in which interacting agents decide their course of action on the basis of a *correct* deliberation. This aspect of the basic premise is reflected in the requirement contained in the premise (d) of the explanatory model.

Taking into account not only these considerations of the extent to which the explanatory model proposed may provide an explanation of the basic premise but also the considerations in this chapter regarding the different conceptions of moral motivation, we may now draw a significant conclusion. We can conclude that the explanatory model introduced above is consistent with three major, mutually exclusive conceptions of moral motivation, but also that the model fits the claims made in the basic premise. This result is significant since it entails, if we apply the model, not only that the resulting explanation is independent of more specific conceptions of moral motivation, but also that the resulting explanation will cover actions of the kind that the basic premise addresses.

Chapter 4

Interaction in cyberspace

As has already been made clear, the aim of this book is to develop a model that will explain the basic premise ([TBP]) with reference to an agent's beliefs concerning the reality of the particular interacting party. This requires some understanding both of the elements constituting interaction and of the notion of cyberspace. The previous chapter elaborated on the notions of intentional action and moral motivation and fitted them into an explanatory model ([Exp]) to be used in the explanation of the basic premise ([TBP]).

In this chapter we will explore some of the concepts used in describing cyberspatial interaction. Some of the conceptual analysis carried out here may seem less relevant to the purpose of our work. By and large the reason for its inclusion will, however, become clear in the following chapter, although some parts of the analysis serve simply to create a common background for reflections on the issues in question.

This chapter also contains another important development. As hinted at in previous chapters, we only take the basic premise to hold for certain specific kinds of interaction in cyberspace. Thus, as part of our investigation of the notion of cyberspace we will also address these specific kinds of interaction in more detail.

T. Ploug, *Ethics in Cyberspace: How Cyberspace May Influence Interpersonal Interaction,*   **65**
© Springer Science+Business Media B.V. 2009

4.1   Cyberspace: Infrastructure and interaction

4.1.1   Conceptual computers and digital electronic machines

To use a tentative and preliminary approximation, a computer may be defined as a device functioning by processing representations of data in accordance with a compiled list of instructions. On the basis of this definition we may move on to consider the conceptual architecture of contemporary computers.

Von Neumann machines

Most modern day computers work on the basis of the stored program architecture also known as the von Neumann Architecture. In this architecture, or conceptual design, a computer is divided into four main interconnected sections. First, there is an arithmetic and logic unit (ALU) capable of performing basic arithmetic and logical operations such as the addition or comparison of numbers. Second, there is a memory unit, which may be represented as a number of cells each having a numbered address. These cells may contain data of two kinds: either instructions on what to do or data on which the instructions are to be applied. An instruction is typically of the sort 'Multiply the content of cell 1 with the content of cell 2, and place the result in cell 3', whereas data may be a string of digits or numbers or the like. Third, there are input and output devices (I/O), which enable the computer to send and receive data to and from external sources. Input devices are, for example, mouse, keyboard and digital camera, whereas output devices might be a monitor, printer and different kinds of sound-generating devices. Fourth, there is a control system with several functions. For example, it both reads data and reads and decodes instructions from the memory or from I/O devices, it provides the ALU with the instructed input and instructions as to the operations to be performed on the relevant input, and it sends the results to the memory or to the I/O devices. Note that the combination of ALU and control circuit is known as the central processing unit (CPU). In personal computers the CPU is placed on a single chip called a microprocessor.[1] Finally, apart from the four main sections the stored programme architecture usually also includes some sort of timer or clock driving the control circuitry. The Von Neumann architecture may thus be represented as in the diagram below:

---

[1] Brookshear in [11, pp. 100–101 and 110–112].

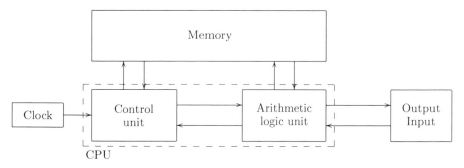

The principle workings of a computer based on the stored programme architecture may be described as follows: The clock initiates a cycle in which the computer retrieves instructions and data from its memory, executes the instructions and stores the results, and then retrieves further instructions and so on. The process continues until terminated by a so-called 'halt' instruction.

Digital computers

Although the stored programme architecture is neutral with regard to several aspects of its implementation, we will focus here on its implementation into digital computers, as modern computers are digital. A computer is classified as digital because of its way of representing instructions and data. In the digital computer data and instructions are represented by a sequence of symbols taken from an alphabet with a fixed number of characters. The alphabet commonly used consists of two characters: i.e., it is binary, and these are typically denoted '0' and '1'. In other words, the digital computer basically functions – processes, transmits, stores, inputs and outputs – by means of the digitization of any other representation of instructions and data: that is, by means of any other representation of instructions and data being constructed as sequence of '0's and '1's. The shortest possible sequence representing other representations of instructions and data, and thus data, clearly consists of either a single '0' or '1'. The placeholder of this, the most basic unit of data, is known as a 'bit'. Related to this is a 'byte', a sequence commonly consisting of eight bits (octet). A byte may consequently represent $2^8 = 256$ different values.[2] In the light of this introduction of the concept of a bit, the process of digitization may be re-described as a process of representing other representations of instructions and data as sequences of bits.

---

[2] Brookshear in [11, pp. 38 and 56–58].

Electronic machines

Having thus considered the conceptual workings of a computer in the Von Neumann architecture and the conceptual meaning of a digital computer, the next step must be to consider the physical or material implementation of these conceptual devices in the modern computers by the use of which we enter cyberspace. We will do this by a very brief examination of both the physics of a computer and its physical appearance.

In terms of the physics of a computer, the simple point to be made here is that the workings of a computer may be defined at a conceptual level as those of processing data in accordance with a set of compiled instructions, where data and instructions are represented by sequences of bits and the processing conforms to the Von Neumann architecture. However, the modern computers with which we are concerned here implement this conceptual design in a technology of the physical or material world. That is, in these computers any of the four sections of the Von Neumann architecture has a physical counterpart, and representations of data and instructions by means of bits also have a physical implementation. To illustrate this, let us briefly look at the microprocessor that makes up the CPU – i.e., the ALU and control system – of a modern computer. The microprocessor is generally made of a solid piece of silicon on which there are circuits consisting of multiple transistors, resistors and the wires connecting them.[3] Similarly to other electrical circuits, the microprocessor operates fundamentally on the basis of charged particles – i.e., the free electrons of the silicon – flowing through the circuit in ways determined by the components and poles of the circuit. This description will suffice to illustrate that the transmission and storage of data and instructions by computers based on microprocessors are processes with corresponding events in the material and physical world. Moreover, we can see that the sequences of bits representing data and instructions have a corresponding representation in the physical and material world.

In terms of physical appearance a computer is usually associated with a set of technological equipment. Thus the term 'computer' is commonly taken to denote the particular piece of technological equipment that enables a person to fulfil tasks of a general purpose, such as word processing, the construction of spreadsheets, multimedia editing, programming, gaming, internet surfing, e-mailing etc. As such, the term 'computer' refers to a combination of electronic equipment (hardware) and programmes (software) which may be accessed by one or more individuals at a time. In the same way, operating a computer will commonly involve sitting in front of a

---

[3] Cf. Borgmann in [9, pp. 155–165] for a detailed account of transistors, resistors and gates.

monitor and using a keyboard and mouse or some other input device to manipulate a cursor in a software-generated environment in a way that qualifies as an activity of the kind mentioned above.

Internet: 'the Net' and 'the Web'

Having clarified the notion of a computer, we will now briefly look at the concept of an 'internet' and other closely related concepts.

An internet is simply a *net*-work of *inter*-connected computers which, by virtue of sharing a protocol, have the ability to communicate and thus to access and share the resources of the other computers in the network. In computing, a protocol is a standard defining how to perform real-time communication: e.g., how a message starts and ends. The internet, 'the Net', is a network of computers communicating in accordance with a particular standard or protocol, the so-called Transmission Control Protocol/Internet Protocol (TCP/IP). As such 'the Net' is an instance of an internet. 'The Net' has to be distinguished from the 'World Wide Web' and e-mail.[4]

The 'World Wide Web', 'the Web', is a service that operates over 'the Net', or, to put it slightly different, it is a resource made available by 'the Net'. Thus in substance 'the Web' is a set of documents and files interconnected by means of hyper-links. A hyper-link is a reference within a document stored on and displayed by a computer that may be used either to retrieve another document or file already stored on a computer, or to dynamically generate a document or file. E-mail may be described as mail composed, sent and received by means of electronic communication systems. Although e-mailing is not logically dependent upon 'the Net', the latter has obviously increased the availability of this service.

4.1.2 Defining cyberspace: Virtuality and interaction

On the basis of the definition of the basic concepts associated with the term 'computer', we can now pursue a tentative definition of the concept of cyberspace. As will become clear, the concept is difficult to comprehend fully, but we will attempt to sketch some of the alleged properties of cyberspace and relate these to the descriptions of computers and internet already made.

As a first approximation, let us say that cyberspace is a virtual place, room or space sustained and accessed through networks of interconnected computers in which

---

[4] See Snyder in [88, pp. 78–82] for definitions of internet, 'the Net' and 'the Web' (see below).

agents are interacting. That is, cyberspace is some sort of place for interaction made available by networks of interconnected computers.

Cyberspace, virtuality and 'the Net'

The first two components of the tentative definition of cyberspace concern the alleged virtual nature of cyberspace and the relation between cyberspace and 'the Net'.

The virtuality of cyberspace refers to its independence of any specific spatiotemporal location. That is, as a place or space of interaction cyberspace does not require the interacting parties to be at one particular location at one particular moment in time in order to meet in this place or space. Interaction in cyberspace is attached to a physical substratum but it may be synchronous or asynchronous, and it may involve agents in almost any conceivable geographical location.[5] Note, very importantly, that the concept of virtuality as defined here is not the opposite of reality, as it is sometimes conceived to be. However, it still encompasses the meaning of *quasi* or *pseudo*, where these are taken to indicate that something is not completely what it seems to be. Thus in being a virtual place or space for interaction, cyberspace is clearly a *quasi-* or a *pseudo*-place for interaction: i.e., it is not a place in the ordinary sense of the word, according to which a place or space for interaction occupies a particular limited spatio-temporal location.

The second important component is the assumed relation between cyberspace and 'the Net'. The tentative definition clearly rejects any possibility either of the concept of cyberspace being merely a synonym for 'the Net' or of it simply denoting the pool of data stored on computers and made available through computer networks. However, as should be clear from the tentative definition, cyberspace is dependent on the existence of an internet. More specifically, cyberspace seems to be a place, room or space that supervenes on the existence and workings of interconnected networks of computers. Hence any change in the state of the relevant interconnected computers, such as the loss of power, will also involve a change in the interaction in cyberspace: for example, the impossibility of interaction.

Cyberspace and interaction

The third important component in the tentative definition of cyberspace is the interpretation of cyberspace as a space for interaction.

---

[5] Levy in [59, pp. 27, 29–30 and 51] and [60, p. 30].

In Chapter 3 we defined an action as being constituted by an agent doing something intentionally that involves bodily movement. In the light of this definition of an action let us define interaction as a binary relation between an agent and a patient constituted by a sequence of reciprocal responses between the agent and the patient in which either the responses of the agent or of the patient or of both qualify as actions. Examples of interaction might be a telephone conversation, playing football, hugging a friend, withdrawing money from a cashpoint, driving a car etc.

Note that by defining interaction on the basis of the relevant concept of action, interaction comes to require the involvement of agents or patients capable of having intentions: i.e., other things being equal, human beings. Hence the definition clearly excludes the effect of the moon on the tides from qualifying as interaction. It is important to note that the definition does not require both the agent and the patient to be capable of (intentional) action, only that at least one of them is so capable. Thus the patient may be any kind of device or entity that lacks the capacity to perform actions but nonetheless may provide some sort of feedback, input or response that may initiate the agent to further actions – and vice versa.

Note also that the concept of response is here used in quite a broad sense. Thus the response of a patient is any kind of reaction to the influence of the agent that is conceived by either the patient or the agent to be a particular reaction to a particular influence: in other words, the particular output from the patient conceived by either the patient or the agent to result from a particular input. Such a response may be anything from an action to a causal effect as long as it is interpreted by either the patient or the agent as the result of a particular influence or input. Again, this definition rules out the possibility of the moon's effect on tides qualifying as interaction. In this case, however, the implication is established by the requirement of either the agent or the patient to interpret the response as a response.

Before considering a few examples of interaction in cyberspace, let us end this brief investigation into the concept of interaction by noting that a sequence of reciprocal responses consists of at least one response, where a response evidently always requires a preceding influence or input conceived by the agent or the patient to be a particular input or influence directed at the patient. In principle a sequence may consist of an arbitrary number of responses. This reflects the fact that interaction may continue indefinitely, and that the responses constituting the interaction may be delivered with varying temporal intervals.

Although this analysis of the concept of interaction is admittedly rather brief, this need not worry us unduly. We are not in this book primarily concerned with drawing conclusions regarding interaction in general, or even with interaction in cyberspace in general. As was mentioned in the introduction, our present work specifically focuses on two particular kinds of interaction in cyberspace, to which we will soon turn.

Examples of interaction in cyberspace

Before narrowing our focus, let us briefly enumerate some of the different ways of interacting in cyberspace.

One such example is internet banking, which allows users, by means of a computer and a specific programme, to access their bank accounts in order to transfer money, pay bills and conduct various other related transactions. Another example is computer gaming. Cyberspace proliferates with opportunities for online computer gaming. In certain kinds of gaming (Multi-User-Dungeon, or MUD) players are represented by a character, called an 'avatar', through which they usually have a first-person perspective on the character's surroundings as they exist in the computer-generated world of the character and are depicted on the screen in front of the player. When playing in cyberspace, the characters of the other players may become part of the world or the surroundings of the other players' characters. Another extensive and interactive use of cyberspace is file-sharing. Several websites offer the opportunity to share files with all manner of content: films, music, pictures, computer programmes, articles, books etc. As with computer gaming, small communities are often built around these sites, as particular groups of people with similar interests gravitate towards them. Cyberspace also offers newsgroups – sites dedicated to the online sharing of news relating to a particular subject. They are functionally similar to discussion forums, which allow people to post messages and questions for discussion in relation to a particular subject and, in the course of time, to receive some kind of response. Cyberspace also promotes interaction in the form of online shopping. It seems as though most companies these days also offer their products in cyberspace. A website displays the products and allows for online payment. If necessary, the product is then delivered to the purchaser's home a few days later. A final example is the widespread interactive use of cyberspace for self-diagnostics. A number of sites offer the opportunity of testing one's physical and mental health, IQ, personality etc. On the basis of filling in a questionnaire presented on the screen of the computer, the site then presents a diagnosis of health, IQ, personality or the like.

In addition to these examples of interaction in cyberspace there is obviously also the use of cyberspace for data search and retrieval. So-called search engines allow users to search for data among the millions of sites representing various organizations, companies, government offices and institutions, networks and communities of people with common interests and so on.

4.1.3   Specific kinds of interaction in cyberspace

There are thus many kinds of cyberspatial interaction. However, we will here limit our focus to the special cases of interaction in chat-rooms and, less explicitly, to telerobotic interaction.

The purpose of this narrowing of focus is to have a few very concrete and well-known examples to which we can apply our reasoning in the following chapter. First and foremost, though, the narrowing of focus is a consequence of the diversity of forms of interaction in cyberspace. Thus, as already mentioned, we do not take the basic premise, [TBP], to apply to all kinds of interaction in cyberspace. Examples of areas outside its application are e-mailing, e-banking and social networking sites such as Facebook and MySpace. E-mailing, e-banking and social networking sites are considered exceptions on grounds that may be made explicit by means of the hypotheses investigated in the following chapter. E-mailing, however, is also left out of consideration because of an important difference between this kind of interaction and the kinds of interaction investigated below. Thus e-mails, so to speak, contain markers of the professional context in which they are often used. A satisfying investigation of interaction involving e-mails would thus have to address the question of the role of the markers of the context: for example, markers of the e-mail being related to a company, organization or the like. Similarly interaction on social networking sites is not characterized by the same level of anonymity as the types of interaction with which we are concerned here. On networking sites such as Facebook and MySpace people typically identify themselves by means of their real names and a portrait simply because of the underlying purpose of being able to connect with friends, relatives, classmates, business connections etc. As will become clearer from the following investigation, anonymity ('hiddenness') is a distinguishing feature of those types of interaction in cyberspace for which we take the basic premise to hold.

Chat-rooms

A chat-room is a forum in cyberspace in which people can communicate by way of realtime broadcasting of messages to other people in the same forum. The chat-room functions on the basis of a web application that allows users to write a message using

their keyboard and post it on a website, where other people may gain access and read
it. This may have an interface as below:

A slightly more advanced kind of chat-room is the so-called three-dimensional chat-
room in which users move around in a three-dimensional world by means of an avatar
and communicate by means of written messages. Second-Life is an example of such
a three-dimensional chat-room. A snapshot of a three-dimensional chat-room – i.e.,
virtual world – can be found below:

The chat-rooms with which we are concerned here have a number of properties that need to be made explicit. First, as already indicated, chat-rooms differ from the newsgroups mentioned above by allowing for real-time or synchronous interaction in the form of communication. The broadcast message is instantly posted on the relevant website, and the response is expected to be received within a short amount of time. Hence the name *chat*-room. Second, although there exist chat-rooms that incorporate audio and video communication, the kind of rooms with which we are primarily concerned only allow for communication on the basis of written messages posted on the website and shown on the screens of the communicating parties. Third, the relevant kind of chat-room requires its users to identify themselves with some kind of name in order to gain access to the chat-room. This name is attached to any message users post in the chat-room. However, the real name and identity of a user may remain unknown both to other users and to those offering the service at a given website. Fourth, the chat-rooms may be accessed and abandoned at any time and free of any charge or other liability.

Quite often websites give access to chat-rooms classified on the basis of subject area. The subjects may cover a wide range of interests, ranging from sexual topics to politics and religion. Sometimes chat-rooms are monitored in order to prevent undesirable behaviour. Thus any user may be excluded from the chat-room if they contravene the etiquette implicitly or explicitly accepted when entering the chat room.

Tele-operation

The second kind of interaction for which we take our considerations in this book to hold is teleoperation.

Etymologically, tele-operation means working with or on something from a distance. In this context the word will be used in a more specific sense: namely, to denote the real-time operation of a device only capable of being accessed and controlled over the Net, and hence by means of a computer. Thus 'distance' here refers to the distance created by the mediation of the operation rather than a geographical distance between the operator and the operated device. That is, the operator and the operated device may be geographically close as long as the operator is capable of accessing and controlling the device only over 'the Net', and thus by means of a computer.

To illustrate the difference between this more specific sense of tele-operation and tele-operation in general, the case of controlling a model aircraft by means of a radio

transmitter is very helpful. Although the controlling of a model aircraft would normally qualify as tele-operation, it does not do so according to the above definition.
This is, first, because the control of the model aircraft is not mediated by a network
of interconnected computers. The aircraft may thus be accessed without the mediation of a computer and the Net. Second, it is because the aircraft is controlled by
means of a radio transmitter and not by means of a computer interconnected with
other computers. It should be borne in mind that we are here using 'computer' in the
sense of a physical machine. However, the controlling of a model aircraft does fulfil
the requirement of taking place in real time – a requirement that is not fulfilled by
other cases of tele-operation, such as the operation of planetary rovers.

To make this narrowing of focus even more specific, let us revert to the example of tele-operation described in the opening chapter. The 'Legal Tender' experiment
described there clearly involved tele-operation. The experiment gave people the opportunity to experiment with $100 bills through a website. Just sitting at their computer
using their keyboard, people were able to control a robot connected to a computer in
another location and perform different kinds of real-time experiments.

A quite similar and very relevant kind of tele-operation was planned to be made
available by a hunting ground in Texas. Simply by using their keyboard, people were
supposed to be able to manipulate a computer-controlled device placed in hunting
grounds and to hunt animals in real time.

It is worth noting that in both these cases the control of the relevant device is based
on visual access to the device and to the location in which the device is to be used.
The latter case even incorporates audio output.

From this point onwards these examples of tele-operation will be considered paradigmatic of the kind of tele-operation with which we are concerned here. Although our considerations in the chapters to follow will mainly be applied to interaction in chat-rooms, the conclusions drawn may readily, it seems, be transferred to these paradigmatic examples of tele-operation.

4.2   Key properties of cyberspatial interaction

On the basis of the definitions and descriptions provided in the previous section we
are now in a position to highlight those properties of interaction in cyberspace crucial
to the reasoning used in the following chapters.

4.2.1   Limited exchange of data and information

The first property of cyberspace to be highlighted in this section is the limitations on
the data exchanged between interconnected computers. The aim here is to establish
that there are *de facto* limits to the amount of data exchanged between computers,
and to show how the limits on the exchange of data influence the availability or
accessibility of information. To fulfil this aim requires some dissection both of the
workings of a computer and of the relationship between data and information.

The limited exchange of data between interconnected computers seems to follow
intuitively from our earlier very brief consideration of the physics of computers. It
was there established that the representation of data in digital computers based on
microprocessors is tied to a physical implementation of the data – that is, to a physical
state of the world. Hence the storage or transmission of data occurs on the basis of
events in the physical world, such as electrons flowing in the circuits of the microprocessor. As such, the transmission and storage of data are conditioned by the possible
limits imposed by the physical world. Without delving into the more fundamental
discussion of the limits of the physical world in terms of transmission and storage of
data, it would be uncontroversial to claim that, given the current way of constructing
computers and networks of computers, there are *de facto* limits to the amount of
data that can be transferred between computers. Such limits are evidenced by the
fact of connections between interconnected computers being classified according to
their bandwidth or throughput, where this is a rate at which the connection is able to
send and receive data. The rate is usually measured in how many bits the connection
is able to transfer per second, and may consequently be termed a 'bit rate'.[6]

Having thus established the *de facto* limitations of computers represented by the
bandwidth of the connection between computers, we will now turn to the task of
uncovering the relationship between data and information, in order to clarify whether
these limits of data are also limits of information. According to what has been termed

---

[6] See Snyder in [88, p. 299] for definition of latency and bandwidth.

the General Definition of Information (GDI), information is the equivalent of objective semantic content. Item σ qualifies as an instance of information if, and only if, σ consists of n data ($n \geq 1$), the data are syntactically well-formed, and the well-formed data are meaningful.[7] Information is, therefore, meaningful data, where 'meaningful' indicates that the data has a meaning and function in a semiotic system, such as a word uttered in a certain language, and where a datum is a relational entity defined as a lack of uniformity between two signs or as the occurrence of a difference: for example, black dots on a white piece of paper.[8]

Regarding the use of meaningful in this definition, it is worth noting that the semiotic system constituting the meaning of the data is usually 'carried', so to speak, by the receiver of the information, but it may also, according to some definitions of information, be 'carried' exclusively by the producer of the data or by the data itself. The Rosetta Stone before it was deciphered, for example, may be considered an example of the former, while the concentric rings in trees indicating their age may be considered an example of the latter. Allowing for data to have meaning independently of the user or receiver of the data introduces an aspect of mind-independence of information, which may be further extended by allowing for information to be independent of both the user and the producer.[9] (GDI) only allows the possibility of mind-independence of the user or the receiver.

In the light of this exposition of the relationship between data and information we may now pose the question of whether or not the data processed by computers qualify as information. Clearly they do. Thus the data represented by sequences of bits and transmitted between interconnected computers are syntactically well formed in terms of satisfying the rules for the construction of permissible, proper or well-formed expressions from certain basic categories of the formal languages of computers. In other words, the data are constructed in accordance with a form of syntax. Moreover, the sequences of bits transmitted between interconnected computers are also meaningful. If one is familiar with the semiotic system of the different formal computer languages, it is possible to understand the meaning of the transmitted data. That is, one may be able to convert the sequences of bits systematically into other, more

---

[7] Floridi in [27, p. 42]. The (GDI) is said to present a concept of data agreed upon by information theorists in recent years.

[8] Floridi in [27, pp. 43 and 45]. Data may be further divided into primary, meta-, operational and derivative data.

[9] A position taken by Dretske in [20], and Borgmann in [34, pp. 96–97].

familiar representations of data, such as words, diagrams, music etc., with a specific semantic content. The upshot of these rather commonplace considerations is that the limits of the transmission of data between interconnected computers are also limits on the transmission of information. Leaving aside the question of how to quantify information, we may interpret this result in rather vague terms as simply establishing that the limited transmission of data between interconnected computers implies that there is a limited transmission of information between them.

As the following chapter will reveal, this result has some important consequences for our ability to form certain beliefs about the interacting party.

4.2.2   Limited sensory access

Where the first property highlighted was the material or physical vehicles that constitute cyberspace, the second key property is related to the activity of human beings in cyberspace. Moreover, it is a property of those specific kinds of interaction in cyberspace listed earlier. Thus in both of these kinds of interaction there is limited sensory access to the interacting party.

In anonymous text-based chat-rooms it seems relevant to distinguish between the patient and the setting, or location, of the patient. In terms of sensory access the agent can see, hear, smell or feel (taste) neither the patient nor the patient's setting or location. Although obvious, this means that the agent does not have sensory access to the properties of the patient's body: e.g., size and shape, scars and handicaps, wrinkles, clothes, hair, skin colour, odour, gestures, facial expressions, decorations, timbre of voice etc. It also means that the agent does not have sensory access to properties such as the colour, temperature, number, shape, size, lightness or darkness of those entities constituting the setting or location, whatever these may be: buildings, nature, furniture, climate, icons or whatever.

Taking the examples of tele-operation described in the previous subsection as paradigmatic, then, it seems relevant to distinguish between the tele-operated entity, the patient ($100 bills or deer) and the setting of the patient. In terms of sensory access, interaction based on tele-operation of the kinds mentioned above enables visual and perhaps even audio access to the operated entity, the patient and the setting and location of the patient. Extrapolating from the examples listed earlier, this will be limited in the sense of involving a very limited number of cameras and microphones, either stationary or movable to a limited degree. This means that there will be a loss of access to those properties of the tele-operated entity, the patient and its setting or location which are only accessible by means of vision or hearing unrestricted in the relevant way. In the case of stationary cameras and microphones being directed at the operated entity and the patient, the properties of the setting may be accessible only to a limited degree or, if movable cameras are employed those properties requiring a unified impression of the operated entity, the patient or setting may be accessible only to a limited degree. These are the properties relative to other properties, such as the relative size of the patient. Moreover, interaction on the basis of tele-operation still restricts the agent from olfactory or tactile access to the operated entity, the patient and the setting as exemplified above.

4.2.3 Extensive anonymity

Like the second key property, the third is also a property related to the activity of human beings in cyberspace. Moreover, it is also a property of those specific kinds of interaction in cyberspace focused on earlier: namely, interaction in text-based chatrooms and interaction based on tele-operation. Thus in both of these kinds of interaction the agent is to a certain extent anonymous.

According to one definition, anonymity is a state in which a person's identity is unknown to others, where identity is the comprehensive set of properties constituting a distinct and unique personality. Examples of these properties are gender, age, temperament, competences, sexuality, education, memories and so on. In this definition anonymity is clearly tied to the whole set of properties constituting the personality, and hence anonymity cannot be a matter of degree – either the set is known or it is unknown. Although it seems as though, other things being equal, a person will always enjoy anonymity according to this definition, it is clearly the case that a person can be more or less close to losing anonymity. Thus a greater or smaller number of the properties constituting a personality can be known by others. The definition of anonymity also refers to the identity and personality of a person. Without going into too much detail, it is worth noting that it is not hereby assumed that identity is some kind of stable, unique and finite set of properties of a person. Although it may be unique and finite, we will assume the relevant set of properties to be dynamic in the sense of being capable of change over the course of time. When referring to the identity of a person or perhaps even their 'true identity', we are thus only referring to the identity or personality of a person at a given point in time – or during a short interval of time.

An alternative and perhaps more viable definition would also have it that anonymity is a state in which a person's identity is unknown to others, but would then define identity as any set of information about a person that would allow others uniquely to identify the person in a group of people. An example of such a set of information would be the personal identification number – also known as a CPR number – issued by the national registration office in Denmark to any citizen of that country. The CPR number uniquely picks out a person from the population as a whole. Like the definition of anonymity proposed above, this one also ties anonymity to the whole set of information allowing the unique identification of an individual. Hence, again, anonymity cannot be a matter of degree. The difference between the two definitions is simply this: in the first definition identity is tied to personality, and in the second definition identity is tied only to the information required to identify a person uniquely among a group of people. However, it seems as though the set of information required to identify a person uniquely among a group of people could possibly consist of a subset of the set of properties constituting a unique personality, without this being a necessary prerequisite.

As mentioned at the beginning of this subsection, both of these kinds of anonymity are found in the particular kinds of interaction with which we are concerned here. Thus in text-based chat-rooms there is no requirement either to reveal properties constituting the user's personality or to give information that would allow others to identify a person uniquely among a relevant group of people. The only public information attached to a person is the freely chosen nickname. As a result, everyone is initially – in other words, before the voluntary revelation of information through text-based communication – anonymous to everyone.

In the case of tele-operation matters are somewhat different in that in both of the examples given – the 'Legal Tender' experiment and the online hunting website – the participants and users have to waive anonymity in the sense of revealing information that will make them uniquely identifiable among other people, notably the suppliers of the websites. They have, for example, to state their name and address. If we suppose that they do this in a verifiable way, they are still anonymous in the very important sense that at least part of their personality remains undisclosed for the suppliers of the website. Since the patients in these examples are not human beings, we will leave the question of their anonymity aside. It is of greater importance, however, that the actual operation of the robot in the 'Legal Tender' experiment or the rifle in the online hunting does not in itself require the surrender of anonymity. Thus there is nothing in the actual controlling of the relevant entities that necessitates a loss of anonymity in any of the given senses of anonymity. Actually it seems as though one may – unless required to do otherwise – maintain a strong degree of anonymity. Contrary to the case of interaction in text-based chat-rooms, one does not in tele-operation of the relevant kinds have to make any statements and thus through the choice of words and their content reveal anything about oneself. The only possible disclosures are those involved in controlling the relevant devices. Thus in principle an agent may to a significant extent remain anonymous, in both possible senses of anonymity, when operating a device in cyberspace.

A final remark has to be made regarding those traces that a computer may leave behind and which may therefore pose a threat to anonymity. A computer is identified on the internet by means of a so-called 'IP address'. Ultimately the IP address may make a singular computer identifiable among the class of interconnected computers, and thus a computer may not be able to remain anonymous in one important sense of anonymity. It is not hard to see that the possible lack of the anonymity of a computer does not entail the loss of anonymity of the individual operating it: by using computers in public places, the anonymity of the individual may be preserved.

With these different senses of anonymity and their application to the examples of interaction in mind we will now briefly elaborate on the logical relationship between the three key properties.

4.2.4 Logical relationship between key properties

A few notes on the logical relationship between the key properties highlighted above must be made, the aim being to show that each of the highlighted properties is separate in the sense of not being equivalent to any of the others. As we will see, this does not mean, however, that they are wholly unrelated.

The limited possibilities of exchanging data and information via computers and networks of computers is clearly not equivalent to the limited sensory access of the agent to the patient in interaction in cyberspace. Although there are limits to the exchange of data and information, this cannot preclude the possibility of sensory access between the agent and patient in interaction in cyberspace. Likewise the limits on the exchange of data and information do not entail the anonymity of the agent or the patient in either sense.

The limited sensory access between the agent and patient is clearly not equivalent to the anonymity of the agent and patient in any of the possible senses of anonymity.

Thus there may be very limited sensory access between the agent and patient without anonymity being preserved, simply because the anonymity may be waived in the communication. On the other hand, extensive sensory access does seem to threaten anonymity. Visual access to the face of another person does, for example, provide information that will make the person identifiable in a group of people. However, the person may still be anonymous in the second important sense of not having their true identity or personality revealed. Furthermore, it does seem as though visual access is the only kind of sensory access that will invariably threaten anonymity in this way.

In order for this analysis to be adequate, a more detailed investigation would have to be conducted. For the present purpose of establishing some kind of independence between the properties, or at least of pointing to the complexity of the relationship between these concepts, our comments in this subsection probably suffice.

Part III

Explaining the basic premise

Chapter 5

Belief and particularity

This, the third part, constitutes the 'heart' of this book. On the basis of the framework provided by our analysis of the basic premise ([TBP]), of our development of the explanatory model ([Exp]) and of our examination of some of the key properties of cyberspace and interaction we will attempt here to provide a model that will explain why human agents may come to act in ethically different ways according to whether they are inside or outside cyberspace.

The debt owed to Emmanuel Levinas and the 'Legal Tender' experiment referred to in the opening chapter will be evident in the explanation provided. Thus we will, in short, explain the basic premise on the basis of considerations of how an agent's belief in the reality of a particular other person may be tied to certain beliefs, and how the absence of 'the face of the other' in certain kinds of interaction in cyberspace may limit the availability of the evidence required to form these underlying beliefs.

Before entering into these investigations, we will begin this chapter by sketching the structure of the analysis and the reasoning of the chapters that make up the third part of the book. Following this, we will set out on the first stage of providing an explanation by introducing three hypotheses linking an agent's belief in the reality of a particular interacting person with certain other beliefs.

5.1 Structure of analysis

Our investigations in this part of the book will proceed in four stages. First, we will present a number of hypotheses for how a particular belief of an agent may be arrived at and, by extension, how this belief may come to depend on underlying beliefs about features or properties of the world. Second, we will proceed to argue how and to what extent the formation of the underlying beliefs may be influenced by the fact of the agent acting in cyberspace or, in other words, owing to the change in the features or properties of the world implied by the agent acting in cyberspace. We will argue, third, that changes in the relevant belief may affect the agent's 'moral motivation'. On the basis of these considerations we will, fourth, provide an explanation of the assumed difference in interaction. To put it schematically, our reasoning in the following sections will have the following logical structure:

[**Stage 1**]
A's belief that p is dependent on the conditions C.

[**Stage 2**]
The conditions C are not satisfied in particular kinds of interaction in cyberspace.

[**Stage 3**]
A's belief that p is a prerequisite of A's motivation to do q, and thus in turn of A doing q.

[**Stage 4**]
Hence A lacks the belief that p in interaction in cyberspace and therefore fails to do q.

As for the more specific content of each of the stages, a few comments may be helpful before entering the analysis. The first stage will see the launch of three hypotheses asserting that an agent's belief in the reality of the particular interacting party, the particular patient, is dependent on other certain other beliefs and on the availability of evidence in support of these underlying beliefs. Such hypotheses introduce a dependency between an agent's belief in the reality of the particular patient and those vehicles that convey evidence in support of the beliefs underlying this belief. It will be argued that it is these vehicles, among other things, that change in the move from interaction outside to interaction inside cyberspace. The first stage is covered in this and the following chapter. In this chapter the hypotheses are briefly introduced and some of the common elements examined. In the next chapter the hypotheses are each given a section in which the key concepts are analysed, the logical relations between the claims of the hypothesis investigated and the central claims discussed on the basis of certain arguments.

The second stage consists in the application of the hypotheses to the case of interaction in cyberspace as exemplified in text-based chat-rooms and tele-operations of certain kinds. This application of the hypotheses will draw heavily on the key properties of the relevant kinds of interaction in cyberspace outlined at the end of Chapter 4. The second stage is covered in Chapter 7.

The third stage is intended to provide an argument to the effect that an agent's belief in the reality of a particular patient may come to influence the behaviour of the agent. More specifically, the argument to be provided here must show not only *that* an agent's belief in the reality of the particular patient may influence the agent's motivation to act in a certain way but also *how* the relevant belief influences the agent's 'moral motivation'. In the fourth stage the threads are drawn together in the attempt to explain the assumed difference in interaction on the basis of the preceding considerations. Both the third and the fourth stage are covered in Chapter 8.

5.2   The three hypotheses

How do we come to be convinced of the reality of a particular other person? The
answer is apparently not as straightforward as with other beliefs. The claim in this
book is that our conviction of the reality of another person is linked to the formation
of certain other beliefs. Moreover, we will here defend the following three hypotheses:

[1. Hyp.]   An agent engaged in interaction with a patient is convinced of the
reality of the particular patient of her actions to the extent that she,
on the basis of what she in the given situation considers to be relevant
and reliable epistemological evidence, comes to believe the patient to
be a determinate entity.

[2. Hyp.]   An agent engaged in interaction with a patient is convinced of the
reality of the particular patient of her actions to the extent that she,
on the basis of what she in the given situation considers to be relevant
and reliable epistemological evidence, comes to believe that certain
of her actions have an effect on the patient, and that the effect of her
actions upon the patient will be constitutive of her life-world.

[3. Hyp.]   An agent engaged in interaction with a patient is convinced of the
reality of the particular patient of her actions to the extent that she,
on the basis of what she in the given situation considers to be relevant and reliable epistemological evidence, comes to believe that the
patient is vulnerable to or dependent on her actions.

The three hypotheses share a number of elements both implicitly and explicitly.

5.2.1   Being convinced to a certain extent

The first element shared by the three hypotheses is the nature of the claims made.
Clearly these hypotheses trade on a distinction between what is real and what is
believed to be real. The advanced claims concern the constituents of our beliefs or
convictions concerning the reality of something and not what constitutes this something being real. In short, the three hypotheses make claims that are epistemological
rather than metaphysical in nature.

In relation to this, the phrase 'to the extent that', which forms part of all the
hypotheses, is to be interpreted as referring to the agent's belief or conviction. In a

generalized form the hypotheses claim that an agent believes in the reality of a particular patient to the extent that the agent comes to believe that P, where P has the character of an indicative sentence. What is meant, therefore, by interpreting the relevant phrase as referring to the agent's belief or conviction is that the agent becomes more certain or convinced of the reality of the particular patient the more the agent becomes certain or convinced of P. This interpretation takes the reality of the particular patient and the content of P to be not susceptible to gradations, and instead implies the gradation of the certainty or conviction with which these beliefs are held. As such, the interpretation is taken to reflect the fact that we generally do hold beliefs with different degrees of certainty or conviction, where the certainty may reflect the degree to which we consider ourselves warranted in holding the relevant beliefs.[1]

It is also worth noting that the hypotheses concern how an agent engaged in interaction arrives at a belief to the effect that the particular patient is real, and not how the agent ought to have formed this belief or what would have made the agent justified in so believing. That is, they concern what is involved in our everyday, pre-philosophical, common-sense dealings with the world, in coming to believe an entity to be real rather than what would be possibly be involved if we were radical philosophical sceptics.

5.2.2 The reality of the patient

Perhaps the most conspicuous, common phrase of the three hypotheses is that of 'the reality of the particular patient'. As indicated in this phrase – and this is extremely important – these hypotheses concern the reality of the patient interpreted not as the existence of the patient as an abstract entity but rather as the existence of the patient as a concrete or particular entity.

The difference between these two interpretations, which is clearly one of degree, becomes clearer when taking their implications into account. Thus if the first interpretation is preferred, then for an agent to be less certain or convinced of the reality of the patient, with whom the agent may, for example, be communicating in a chatroom, simply means that that agent is less certain or convinced of the reality of the interacting part as an instance of human being *simpliciter*. If, on the contrary, the second interpretation is favoured, then for an agent to be less certain or convinced of

---

[1] Cf. [15, pp. 61–64] for possible definitions of certainty.

the reality of the patient, with whom the agent may, for example, be communicating in a chat-room, simply means that that agent is less certain or convinced of the reality of the interacting part as a *particular* entity. Hence in the second interpretation the agent may come to be less certain or convinced of the reality of the patient as a person with certain characteristics without being less certain or convinced of the reality of the patient being a person.

The choice of the second interpretation is, to argue negatively, based on the first reading being counter-intuitive. It simply does not seem reasonable to suppose that an agent would at any point in, for example, interaction in a chat-room doubt the existence of the patient as someone. To doubt this would seem reasonable if, for example, chat-rooms were also inhabited by computers indistinguishable from human beings in their ability to communicate, in other words if chat-rooms were also inhabited by computers being able to pass the Turing test.[2]

The choice of the second interpretation is also, to argue positively, based on the following understanding of a situation of interaction. In interacting in, for example, a chat-room an agent will on the basis of the available evidence, other things being equal, form beliefs regarding the properties of the patient. That is, an agent engaged in interaction will, other things being equal, form beliefs in which the patient is taken to have the existence of an entity characterized by certain properties ascribed to the patient on the basis of the available evidence. The patient is, so to speak, believed to be the instantiation of a set of properties, where these properties are ascribed to the patient on the basis of the available evidence. Thus an agent communicating with a patient in a chat-room may in the fairly early stages of the conversation believe the patient to be an older woman trying to find someone to engage in chat on the basis of the unsolicited statement from the patient that 'You certainly do seem to be a most agreeable and handsome young man'. In believing this, the agent is seemingly believing the patient to be the instantiation of a (rather limited) set of properties, namely those of being a woman, older, educated and having the intention of engaging in chat – and clearly (more or less present) those generic properties implied by the particular properties respectively, such as the patient being a human, a person etc.[3]

---

[2] Cf. Turing in [96].

[3] We do not take certain genus–species relations between properties to generate problems of opacity for an agents beliefs in the properties of the patient. That is, we take it to be a truth-preserving inference that if an agent believes the patient to be a woman, then the agent also believes the patient to be a human being or a person.

As the interaction progresses, the agent may obviously form more and more complex beliefs about the properties of the patient by virtue of further evidence becoming available.

From this account of the situation of interaction in a chat-room it seems to follow that an agent may at any point in the interaction come to be less certain or convinced of the reality of the patient in the sense that the agent may come to doubt that the patient possesses the set of properties ascribed to the patient at a prior point in the interaction on the basis of the available evidence. That is, the agent may come to be less convinced of the patient existing as someone embodying that particular set of properties which was ascribed to the patient at a prior point in the interaction. To put it slightly differently, the agent may come to be less convinced that the description of the patient denotes an existing person and not a mere fiction – without doubting, however, the existence of the patient *simpliciter*.

Given that the notion of an agent being less certain or convinced of the reality of the patient is to be understood as suggested here, it is worth noting the striking parallel between this situation and that of interacting with an actor in a play. Interacting with another person in a play shares exactly this feature: namely, that an agent may question the reality of a particular set of properties that evidence has led to believe constitutes the patient – without, however, doubting the existence of the interacting party as a person simpliciter. That is, an agent interacting with another person in a play may question whether the properties ascribed to the other person in the course of the play on the basis of the available evidence truly are properties of that other person – or whether these are fictitious properties in the sense of belonging to the character played by the interacting person.

Given that the notion of an agent being less certain or convinced of the reality of the patient is to be understood as suggested here, it is also worth noting how exactly this feeds into the overall interpretation of the three hypotheses. All three hypotheses are to be read as making a claim to the effect that an agent's conviction concerning the reality of a particular patient is tied to having evidence supporting certain other beliefs about the patient. The agent may become convinced that the set of properties describes an existing person insofar as the agent comes to hold these beliefs about a set of properties ascribed to the interacting party. More specifically, if an agent takes the interacting party to be characterized by the set of properties $p_1, p_2 \ldots p_n \subseteq P$, then it is insofar as the agent comes to hold certain beliefs about that set of properties $p_1, p_2 \ldots p_n \subseteq P$ – that they describe a determinate, causally affected and vulnerable entity – that the agent may come to be convinced of the existence of the other person as a person characterized by the set of properties $p_1, p_2 \ldots p_n \subseteq P$.

In this reading an agents belief in the existence of the patient as a person characterised by the set of properties $p_1, p_2 \ldots p_n \subseteq P$ is taken to be a conjecture or hypothesis, which may be shown to hold if the agent receives evidence in favour of the patient being a determinate, causally affected and vulnerable person and the more relevant and reliable evidence in favour of these beliefs that is available to the agent, the more the agent will be convinced of the reality of the patient constituted by the set of properties $p_1, p_2 \ldots p_n \subseteq P$. Hence the real outcome of these hypotheses is their claim to the effect that, in coming to be convinced of the reality of another person constituted by a certain set of properties, evidence is not simply evidence, nor beliefs simply beliefs. Certain evidence plays a more decisive role because it may support certain beliefs in playing their own crucial role in our conviction that the person is constituted by the set of properties in question. This is important to bear in mind.

5.2.3   Reliable and relevant evidence

As a final common feature to be addressed in this chapter, we shall deal with the notion of 'relevant and reliable epistemological evidence'.

Evidence

Let us start by considering what may constitute sources of evidence of a patient's properties. Most often, evidence refers to a set of beliefs based on observations that play the role of supporting a hypothesis.[4] Such evidence is related to perception since perception may be defined as the epistemic process of coming to believe in the instantiation of a particular thing (or fact) on the basis of accessing the thing by means of one's external senses. Hence perception differs from merely seeing, smelling, tasting, touching or hearing a thing by involving the formation of a belief identifying the things seen, smelled, tasted, touched or heard as a particular thing.[5] To summarize, perception of things involves the formation of observational beliefs that may count as support for a given hypothesis. That is, perception is a possible source of evidence. Moreover, it is the only relevant source of evidence in this context.

---

[4] Cf. [47, p. 254].

[5] Cf. [47, p. 652].

As a source of evidence perception may clearly be indirect. Examples of this are the perception of things through images made accessible to us through various sorts of media such as screens or displays (computer, television, film), books, radio and so on. On the contrary, it is not clear that perception may ever be direct. Thus it has been claimed that perception always involves a so-called 'sense-datum', which is a subjective mental representation of an external object that has the properties the external object appears to have.[6] Insofar as this is the case, it clearly follows that direct perception is not possible.

In accordance with our elaboration of the basic premise, [TBP], we will distinguish between face-to-face perception and perception that is not face-to-face. In this context the distinction is to be understood in such a way as to imply that perception is not face-to-face if the perceptual, and hence informational, access to the perceived is *significantly* reduced because either the perceiver and the perceived are spatio-temporally distant or because the perception is mediated by a material object existing independently of the perceiver. Two important remarks have to be made regarding this tentative circumscription of face-to-face perception. First, the causes of reduced perceptual access, as a matter of contingency, seem to be co-extensive in a lot of situations apart from the one situation in which the perceiver and the perceived are temporally close but spatially distant, and the perception does not involve a material object existing independently of the perceiver. Thus, it seems as though temporally distant people can perceive each other only through a medium of the relevant kind, while people who perceive each other through a medium of the relevant kind are in most cases temporally or spatially distant. (Note that there are obvious examples of this not being the case involving e-mail, text messages, phone etc. However, the claim is only that this co-extensionality holds in a large number of cases.) The second point is that we need not trouble ourselves about finding a fully adequate definition of face-to-face perception since we are first and foremost intending to draw conclusions not about face-to-face interaction and perception in general, but only about a particular kind of interaction and perception that is not face-to-face by virtue of being mediated by computers.

---

[6] Cf. [47, p. 822].

Reliable evidence

Having thus elaborated on the concept of evidence and its link to perception, let us consider what is to be understood by the notion of reliable evidence employed in the statement of the three hypotheses.

The three hypotheses employ the concept of reliable evidence in stating the claim that an agent uses evidence taken to be reliable in forming the different beliefs underlying the belief in the reality of a patient. What is here meant by 'reliable' or 'reliability'? The concept of reliability is here used in the sense according to which the reliability of evidence is a matter of the evidence tracking the truth, where truth is taken to consist in correspondence between a truth-bearer and a truth-maker, usually taken to be a proposition and a fact respectively. Thus a piece of evidence is reliable to the extent that it corresponds to facts. Consequently a piece of evidence is unreliable if it does not correspond to facts.

In this attempt to clarify the concept of reliable evidence it is also worth noting that the hypotheses seem to suggest that a consideration of the reliability of a piece of evidence is dependent on a particular agent's consideration of the situation. This seems a plausible suggestion. Thus it seems as though a consideration of the reliability of a piece of evidence is sensitive both to an agent's beliefs and to the context. Imagine an agent of whom it holds, first, that she is at work, second, that she believes that her father would under no circumstances call her while at work, and third, that she is passed the message that her father is on the phone. In this setting the agent will, other things being equal, not consider the message to be reliable evidence for the belief that her father is on the phone. Imagine now the setting is slightly changed, so that the agent instead holds the belief that her father would call her at work in the case of something important coming up. In this modified setting the agent will, other things being equal, consider the message reliable evidence for the belief that her father is on the phone. Imagine now, as a third possibility, the original setting being changed such that the agent is visiting her grandparents instead of being at work. In this setting the agent will clearly also, other things being equal, consider the message reliable evidence for the belief that her father is on the phone. Hence it seems as though we may conclude that there is the professed dependency between an agent's consideration of the reliability of a piece of evidence on the one hand and the beliefs and the context on the other.

Relevant evidence

The three hypotheses also employ the concept of relevant evidence in stating the claim that an agent uses evidence taken to be relevant in forming the different beliefs underlying the belief in reality of the patient. What is meant here by relevant evidence? The concept of relevance is here used in the sense according to which a piece of evidence is relevant to the extent that it may be used to support or justify a given belief. Irrelevant evidence is consequently evidence that may not at all serve as support or justification for a given belief.

In terms of an agent's consideration of the relevance of a piece of evidence, the hypotheses also suggest this to be dependent on the agent's beliefs or knowledge as well as on the context. As in the case of the reliability of a piece of evidence, this seems a plausible suggestion. Thus it seems as though a consideration of the relevance of a piece of evidence is sensitive to an agent's beliefs as well as to the context. Imagine an agent of whom it holds: first, that she is at home in her apartment wondering whether the creature running in the neighbouring apartment a short while ago was two- or four-legged; second, she believes that the human body has 205 bones; and third, she is told by her enigmatic boyfriend, having peeped from the balcony into the adjacent apartment, that the creature has approximately 200 bones. In this setting the agent will, other things being equal, consider the testimony of her boyfriend relevant evidence for her belief that the creature was a human being with two legs. Imagine now the setting is slightly changed so that the agent does not believe the human body to have 205 bones. In this setting the agent may, other things being equal, just as well consider the testimony to be evidence for the belief that there was a dog, a cat or an elephant running around in the adjacent apartment, and hence irrelevant to her belief that the creature was a human being with two legs. Imagine now, as a third possibility, the original setting being changed so that the agent is participating as a coroner in the autopsy of those killed during the genocide in Rwanda and is particularly interested in knowing whether the body, the remains of which are lying on the table in front of her, belonged to a victim with only one or both legs. She still believes the human body to have 205 bones, and she is told by a colleague, having done the initial examination, that the body in front of them has approximately 200 bones. In this setting, where the context is changed but the belief and testimony remain the same, the agent will, other things being equal, not consider the testimony relevant evidence for her belief that the creature was a human being with two legs since it may just as well as be evidence in favour of the belief that the victim had only one leg.

On the basis of these reflections on the concept of relevant evidence it seems as though we may conclude that the outcome of an agent's consideration of the relevance of a piece of evidence is dependent both on the agent's beliefs and on the context, where the context encompasses both the external circumstances and the agent's epistemological interests: for example, the agent's interest in deciding whether the creature running was a two-legged human being or the interest in deciding whether the victim had only one leg.

Chapter 6

Belief and reality

The previous chapter saw the introduction of the three hypotheses central to our attempt to provide a model that will explain the basic premise ([TBP]). The three hypotheses all link an agent's conviction concerning the reality of a particular agent to certain other beliefs and to the availability of evidence in support of these other beliefs. As has already been emphasized, it is vital to understand that the hypotheses are epistemological and not metaphysical in nature, and also that the hypotheses deal with an agent's conviction concerning the reality of a particular other person and not the reality of an abstract entity.

In this chapter we will examine these hypotheses further. The overall aim is to show that our conviction concerning the reality of a particular other person with whom we are interacting is indeed composite – an epistemological construct influenced and logically dependent on the formation of other beliefs. The more specific aim is to show that our conviction concerning the reality of a particular other person is dependent on those specific beliefs captured in the three hypotheses. In doing this, we will have covered important ground in accounting for the role of 'the face of the other' for the ethical character of interaction.

T. Ploug, *Ethics in Cyberspace: How Cyberspace May Influence Interpersonal Interaction,* **99**
© Springer Science+Business Media B.V. 2009

6.1 Hypothesis I: Reality and determinateness

Briefly put, the first of the hypotheses claims that our sense of the reality of particular agents and events is linked to our beliefs in the determinateness of those agents and events. In this section this claim is investigated further.

The investigation has two parts, which are reflected in the overall structure of the section: on the one hand an analysis of the notion of determinateness, and on the other an analysis and discussion of the claimed relation posited between our sense of reality and the belief in the determinateness of the interacting party.

6.1.1 Determinateness and determinedness

For the sake of clarity and focus let us start by restating the first hypothesis and apply it to the special case of the patient being another human being. The first of the hypotheses states that:

[1. Hyp.]   An agent engaged in interaction with a patient is convinced of the reality of the particular patient of her actions to the extent that she, on the basis of what she in the given situation considers to be relevant and reliable epistemological evidence, comes to believe the patient to be a determinate entity.

In the special case of the patient being another human being or person, the hypothesis simply claims, to put it rather simplistically, that the strength of our belief in the reality of the particular person with whom we are interacting is influenced or constituted by the person appearing to us as one person rather than another: in other words, by the person appearing to us as a unique identity with a specific history, personal characteristics such as gender, age, opinions, emotions, values and so on.

Determinateness

The notion of determinateness plays a crucial role in the hypothesis as set out above. In this section we will explore the concepts of determinateness and determinedness in order to enable a clear understanding of what is here meant by an agent coming to believe in the determinateness of the patient. We are not hereby assuming that this belief reflects an awareness of a theoretically constructed concept of determinateness, as contrasted with a similarly constructed concept of determinedness, but rather that

these concepts may be helpful in simply defining the state of affairs believed to have
obtained when an agent believes a patient to be determinate.

Let us preliminarily define determinateness as follows:

[Da]   An event occurring at time t is determinate at the time t' if, and only
       if, it at t' holds for any property whether or not the event possesses
       the property at t.

As such, determinateness may be contrasted with determinedness, which may be
defined as follows:

[De]   The occurrence or non-occurrence of an event at the time t is deter-
       mined at time t', if, and only if, it at time t', as a consequence of
       natural laws and initial conditions, is given which relevant properties
       the event will have at t.

These preliminary definitions capture the most important aspects of determinateness.[1]
First of all [Da] catches the very basic intuition that determinateness is a matter of
all the properties of an entity or event being settled, where the properties of an entity
or event have been settled if, and only if, it has been decided for any given property
whether or not the entity or event has it.

The second important aspect of determinateness captured by [Da] is perhaps less
obvious. However, given that the temporal location of an entity is not a property of
that entity – but rather, like existence, is a property of the properties of the entity[2] –
then by saying that an entity is determinate in the sense of it holding for any given
property that either the entity has it or not, nothing has been said in terms of the time
at which the entity will have or not have a given property. Thus any specification of
the determinateness of an event or entity must have explicit temporal reference points
in order to be adequate. That is, any specification of the determinateness of an entity
must refer to a time at which it holds that the entity either had, has or will have
a certain property or not – in other words it must refer to a time t', at which it
holds for an entity and any given property that the entity either has it or not at a

---

[1] The definitions are taken from Ploug in [80, p. 123]. Cf. also Dorato in [19, pp. 28–29]. For
definitions of natural law and of the fixing-relation between natural laws and the state of the
world see Dorato [19, pp. 80–83 and 86–89].

[2] We are here following Kant, Frege and Russell in treating existence as a second-order predicate.
Cf. [47, p. 257].

time t. Hence the inclusion of these temporal points of reference in the definition of determinateness, [Da].[3]

It is worth noting that not only does the inclusion of temporal points of reference in the definition of determinateness solve a problem of adequacy, but it also makes clear and explicit the possibility of holding different ontological positions on the determinateness of entities or events in the past, present and future. Hence one may, for instance, claim past events to be determinate by holding the position that for all entities and events occurring at a time t they are determinate at time t' if, and only if, t' is later than t. (This is examined in more detail in the discussion of the hypothesis below.)

Third, by contrasting determinateness, [Da], with determinedness, [De], it also becomes clear that determinateness is a logical property of an event or entity in the sense that it is not a property that by definition is fixed by natural laws and initial conditions – or for that matter a property entailed by the event, or entity, being determined, since determinedness by definition fixes not all the properties of an event or entity but only those relevant for the occurrence or non-occurrence of the event or entity. Thus there are no *prima facie* implications between the determinateness and the determinedness of an event or entity – the determinateness of an event or entity is a wholly different property of that event or entity from its being determined.[4]

6.1.2   Belief, reality and determinateness

The first hypothesis basically makes two claims. The first, not surprisingly, is that our belief in the reality of a particular patient with whom or which we are interacting is linked to a belief in the determinateness of the entity. The second is that our belief in the determinateness of the patient is linked to us having some kind of evidence of the determinateness of the patient.

Since there seems to be no *prima facie* reason to suppose that the belief in the determinateness of the patient is formed any differently from the formation of other beliefs about the patient, the second claim may be taken to assert that we generally form beliefs about the patient on the basis of what we consider to be relevant and reliable evidence. As such, the second claim is shared with the second and third hypothesis, and therefore will be dealt with in the final section of this chapter.

---

[3] Cf. also Ploug in [80, pp. 124–125].

[4] Dorato in [19, p. 28] and Ploug in [80, p. 124].

The remainder of this section is dedicated to the analysis and discussion of the first claim. The analysis and discussion falls into two parts. First, the logical character of the first claim of the hypothesis is uncovered, and then an attempt is made to support the hypothesis in and through the consideration of a fictitious case.

Believing in determinateness

The first claim of the hypothesis – that we are convinced of the reality of a particular patient to the extent that we come to believe the patient to be determinate – is to be understood in a rather weak sense. To be more specific, the hypothesis is to be understood as claiming only that the determinateness of the patient plays an important role in the strength of the agent's belief in the reality of the particular patient. Consequently, the belief in the determinateness of the patient is not thought to be a necessary or sufficient condition for the agent being convinced of the reality of the particular patient. It follows that it is fully consistent with the first central claim of the hypothesis both that an agent comes to be convinced of the reality of a patient without believing the patient to be fully determinate, and that an agent believes a patient to be fully determinate without believing the particular patient to be real. Consequently the hypothesis allows for an agent to believe in the reality of quantum physical entities and events, although they are indeterminate (and undetermined), but also for someone to believe in the determinateness of possible worlds as defined within the field of modal logic.

A second point of clarification also has to be made: namely, that the definition of determinateness provided above lacks temporal specification. Hence it has to be specified whether it is our belief in the determinateness of the entity or event at all times that is linked to our belief in the reality of the relevant entity or event or perhaps only a belief in the determinateness of an entity or event at particular times. It seems as though a basic intuition of most people is that past and present events or entities are determinate, whereas future events or entities are indeterminate. This reflects the belief that the past and present simply cannot be any different from what they are, whereas the future is open in the sense of containing only possibilities. If this is truly our intuition, then it clearly follows that the determinateness linked to our belief in the reality of a particular event or entity, such as a particular patient, is the determinateness of the event or entity in past and present time. The first claim of the hypothesis may consequently be rephrased to state that the strength of our belief in the reality of a particular patient is linked to a belief in the determinateness of the particular patient in the past and present, in other words to a belief in it holding at time t for any property that the entity either has it or not at time t', where t' is no later than t.

Note that in our endeavour to substantiate the first of the claims of the hypothesis it is only important that our intuition has the content asserted above. Philosophical investigation may reveal that we have no good reason to consider the future more open than the past or present. However, since we are here concerned with the phenomenology of coming to believe in the reality of a particular patient, what really matters is that our basic intuition supports the indeterminateness of the future alone. Having thus elaborated on the content and logical nature of the first claim of the hypothesis, let us now turn to the task of grounding it. In doing this, we will introduce the so-called 'joker argument'.

The joker argument

As the heading suggests, the main argument to be advanced in this paragraph involves the situation of someone trying to pull another person's leg. Before launching this argument let us briefly consider the *prima facie* credibility of the first claim of the hypothesis.

On the face of it, the claim seems quite reasonable, in that it seems as though we are in general considering determinateness to be an important marker of someone being real as a particular someone. Consider, for instance, soap operas – those open-ended, dramatic and fictitious stories mainly broadcast on TV, in which a group of almost ordinary people in an almost ordinary setting constantly develop new relationships of a romantic or sexual character or become the victims of the most remarkable and unbelievable coincidences. It seems true to say that we generally do not believe the characters in soaps to be real, and that at least part of the reason for this is our belief that the characters in them are indeterminate, whereas real characters are determinate in terms of their past and present. That is, we believe the particular characters to be unreal simply because we believe that they, contrary to real characters, are in a certain sense incomplete – i.e., simply because we believe that it does not hold for any proposition regarding the past or present state of the particular character that it is either true or false. To put it popularly, the particular characters are believed to be unreal because we believe that there are not answers to all questions regarding the past and present of any of them, such as answers to the questions of whether or not this character woke up on her fourth birthday lying on her back, or whether or not that character used a pencil on his first day in school or whether or not a third had a headache after her first swim and so on. 'Characters in a soap opera may be more or less shallow, but there is always a level below which there is nothing.'[5]

It is important to note that this observation regarding our association of the determinateness and the reality of a particular entity is quantitative in nature, in the sense that it does not distinguish between the content of the properties ascribed to a given entity. Obviously a character in a soap opera may also be believed to be unreal on the grounds that it is ascribed properties that we somehow do not believe to be compatible with other of the character's properties: for example, a character who supposedly possesses the property of being human cannot also possess the property of being able to fly or the property of never ageing. However, it is not the relationship between our sense of reality and the content of properties ascribed to a given entity that is under investigation here, but only how the richness of properties of a given entity may influence our belief in the reality of a particular entity through the formation of a belief in the determinateness of that entity.

Despite the apparent credibility of the first claim in the light of our interpretation of the beliefs surrounding soap operas, we cannot be said to have vindicated the claim that our belief in the determinateness of an entity is important for our belief in the reality of a particular entity. In the attempt to do this we will here develop the argument referred to as the 'joker argument'. Imagine two people, A and B, sitting in a caf and having a conversation about a third person, C. Person A is vividly reporting to B how in the preceding weeks he has been dating the most amazing and inspiring woman, C, who he now believes may be the love of his life. For some reason the story raises B's suspicion regarding the reality of the events described. That is, he comes to suspect that A is pulling his leg. As a consequence B starts asking A all sorts of questions about the woman C: for example, about her age, values, job, looks, education, family, religious and political and organizational affiliations. Person A answers these questions.

Before making explicit what we take this case to show, a few interpretative comments may be helpful. First of all, the situation of two people having a conversation is chosen because it seems to be one of those events in our everyday lives in which

---

[5] Borgmann in [34, pp. 99–100]. In contrast to Borgmann, our argument seeks to establish a relation between our sense of the reality of a particular patient and determinateness. Furthermore, Borgmann employs a different concept of determinateness, according to which a token is determinate if, and only if, it possesses all the properties of its type.

we may come to question the truth of the statement of others, and hence the reality of the events described, since it seems as though truth is intuitively taken to
consist in correspondence between a proposition and reality. Second, we take it that
B's questioning of A may be grounded in a number of motives. Thus, part of the
reason for this questioning may be to catch A out through inconsistencies regarding
the properties of C simply because we believe such inconsistencies to be a possible
indication that the story is a concoction of an imaginative, though inconsistent, mind.
The questioning may also be conducted in order to track any hesitation in answers
to ordinary questions regarding C simply because we believe such hesitations to be
a possible sign of creative intellectual power at work. Furthermore, the questioning
may also be conducted in order to decide whether A is able to supply an amount of
detailed information we find incompatible with our beliefs regarding what a human
being both should and may be able to perceive in a situation of the relevant kind and
subsequently remember.

However, the case of the possible joker may come to count in favour of the first
claim of the hypothesis on the basis of three suppositions all enjoying a certain prima
facie plausibility. First, it seems plausible to suppose that B may come to believe
the statements made by A to be true solely on the basis of their conversation, and
hence to believe in the reality of the particular events and entities described. Second,
it seems plausible to suppose that B may come to believe A's story to be true, and
hence to believe in the reality of the particular events and entities described, for a
number of reasons, among which may be the absence of inconsistencies, hesitation
and signs of perceptive and cognitive powers below or above expectations. Third, it
seems plausible to suppose that B may also question A partly because he assesses the
credibility of A's story on the basis of the amount of information A is able to provide.
Consequently it seems plausible to suppose that B may also come to believe A's story
to be true, and hence to believe in the reality of the particular events and entities
described, partly as a result of A's ability to provide vast amounts of information
about C – clearly on condition that the vast amount of information does not come to
count against the truth of the story for other reasons. If we put ourselves in B's shoes,
then, it seems as though the amount of information provided by A may be indicative
of whether or not the story originates in real events and entities such that, the more
(even contextually irrelevant) information A is able to provide, the more we would
tend to believe the information to be reliable and true, and hence believe in the reality
of the particular events and entities described, insofar as the quantity of information

does not come to count against the reality of the entity for other reasons. Fourth and finally, it seems plausible to suppose that B may come to believe A's story to be true, and hence to believe in the reality of the particular events and entities described, because the vast amount of information about C provided by A may count as evidence in support of B's belief in the determinateness of C. Again, putting ourselves in the shoes of person B, it seems as though the more information we have about certain events and entities described by A, the more we would tend to believe in the reality of the particular events and entities, simply because the information comes to count as evidence in favour of our belief in the determinateness of these entities and events. That is, the more information we have, the more we tend to believe that this is an entity for which it holds that for any property it will either be the case that the entity has the property or not at a particular point in the past or the present, and the more we tend to hold this belief in the determinateness of an entity, the more we tend to believe the particular entity to be real – as long as the vast amounts of information do not come to count against the reality of the particular entity for other reasons.

In terms of supporting the first claim of the hypothesis, the four suppositions have clearly served their purpose. Thus the four suppositions regarding the situation of A having a conversation with B on the subject of C point to the conclusion that believing in the determinateness of a person is important for the formation of a belief in the reality of this particular person. Given that B questions A in part because he assesses the credibility of A's story on the basis of the amount of information A is able to provide such that the more information A is able to provide the more likely B is to believe in the reality of C, then it seems plausible to suppose that the information about C counts as evidence in favour of the determinateness of C, and the belief in the determinateness of C is then taken to be the mark of the reality of the particular person C.

Let us end this subsection by considering the strength of the argument launched. The case of the possible joker was introduced as an attempt to vindicate the claim that the belief in the determinateness of an entity is important for our belief in the reality of the relevant entity. The argument proceeded by way of sketching a situation that was interpreted along the lines of a number of seemingly plausible suppositions and ended by concluding that we seem to associate the determinateness of an entity with the reality of the entity. The argument may clearly be countered on the grounds that the sketched situation of A and B having a conversation about C does not in itself provide much support for the plausibility of the suppositions on which the conclusion rests. The joker argument, it may be held, seems close to simply begging the question. Our answer to this accusation is the following: the interpretation of the case of the possible joker is admittedly somewhat suggestive. It seems to gain plausibility, however, if one positions oneself in B's shoes – and it is exactly this exercise we have tried to perform throughout this paragraph.

6.2   Hypothesis II: Reality, causality and life-world

As may be remembered, the second of the hypotheses posited in the previous chapter claims, briefly, that our belief in the reality of a particular other person is linked to our beliefs regarding our own causal effectiveness. In this section this claim is investigated further. On the basis of a rather tentative concept of causation, the investigation will focus on grounding the claim that our sense of the reality of a particular other person is tied in both to our belief in being able to affect the patient causally and to our belief in the causal effect on the patient being formative for our own life.

The investigation is tripartite, and this is reflected in the overall structure of the section. First the notions of causation and life-world are analysed. Then follow an analysis and discussion of the relation proposed between our sense of reality of a particular other person and our causal effectiveness. Finally, the asserted relation between our sense of reality of a particular person and our life-world is analysed and discussed.

6.2.1   Causality and life-world

For the sake of clarity and focus let us, once again, start by restating the second hypothesis and apply it to the special case of the patient being another human being. The second of the hypotheses states that:

[**2. Hyp.**]   An agent engaged in interaction with a patient is convinced of the reality of the particular patient of her actions to the extent that she, on the basis of what she in the given situation considers to be relevant and reliable epistemological evidence, comes to believe that certain of her actions have an effect on the patient, and that the effect of her actions on the patient will be constitutive of her life-world.

In the special case of the patient being another human being, the hypothesis merely claims, to put it rather simplistically, that the strength of our belief in the reality of the particular person with whom we are interacting is constituted in part by the patient appearing to us as affected by our actions – for example, that the person is harmed by certain words or deeds – and in part by the effects of our actions on the patient appearing to us as also influencing our own life – for example, that the other person rejects or relinquishes our friendship. In slightly other words, we believe the particular someone with whom we are interacting to be real to the extent that we come to believe, on the basis of being confronted with the interacting party, that we have causally affected the interacting party and that the effect on the interacting party will influence our own life.

Causation

The hypothesis clearly trades on the concept of causation. Thus it speaks of an agent coming to believe that her actions are having a particular and recognizable effect on the patient. We will now explore some of the possible conceptions of causation in order to enable a clear understanding of what is here meant by an agent coming to believe herself to be the cause of a certain effect on the patient. As with the concept of determinateness investigated in the previous section, we are not assuming here that an agent's belief in being the cause of a certain effect reflects an awareness of a particular theoretical conception of causation, but rather that these conceptions may be helpful in shedding some light on the relationship believed to hold between certain entities when someone believes something to be the cause of something else. As the sole purpose is to clarify the content of the hypothesis, we need not worry that the following account covers only two conceptions of causation – namely, regularity theories and manipulability theories – and therefore may be claimed to be less than exhaustive. It is of greater importance that these conceptions are adequate for the purpose of explicating the claim of the hypothesis.

As the name suggests, regularity theories define causation in terms of an invariable pattern of succession holding between something qualifying as a cause and something else qualifying as an effect.[6] At the heart of this approach to causation clearly lies an assertion to the effect that the occurrence of an entity qualifying as the cause necessitates the occurrence of the entity qualifying as the effect, and hence, briefly, that the occurrence of the cause is sufficient for the occurrence of the effect – otherwise there would simply not be an invariable pattern of succession. Given this general understanding of the relationship between cause and effect, the distinguishing factor between regularity theories becomes the component of the theory specifying which entities may qualify as a cause.

A fairly influential regularity theory defines a cause as an object followed by another, where all objects similar to the first are followed by objects similar to the second.[7] As is evident, this definition allows any object to qualify as a cause insofar

---

[6] Hitchcock in [45, p. 2].

[7] Hume in [49, pp. 60–61].

as it is related to another object in an invariable pattern of succession. This rather inclusive component of the theory makes it suffer from two significant drawbacks. The first is illustrated by the existence of accidental regularities such as the flashing light or beeping sound indicating the arrival of an elevator at its destination floor always being followed by the opening of the door. Although there is regularity here in the sense of an object invariably following on from another object, it is clearly not an instance of causation, as we would not take the light or the sound to be the cause of the opening of the door.[8] The second is illustrated by the existence of imperfect regularities such as, for example, the occurrence of smoking without the development of lung cancer. Although we would accept smoking to be the cause of lung cancer, there is clearly no regularity in the sense of the one object invariably following on the other.[9]

Before moving on to consider an alternative regularity theory, let us briefly mention an attractive feature of this theory. Although the theory may not be altogether viable, the implied epistemology of causation counts in favour of it. Thus, this theory is able in a rather simple manner to account for the way in which we come to know that p causes q: we simply observe that $p's$ are invariably followed by $q's$.[10]

A more refined regularity theory holds that the regularity constitutive of causation is a relationship holding between the cause defined as an 'inus' condition and the effect, where an 'inus' condition is an insufficient but necessary part of a complex condition which is unnecessary but exclusively sufficient for the occurrence of the effect.[11] Suppose that a fire in a house is claimed to have been caused by a short circuit. The meaning of this claim can be neither that the short circuit was necessary nor that it was sufficient for the fire to occur, as the overturning of a lit candle would have been sufficient, and it could not have occurred without flammable materials, oxygen etc. It means that the short circuit was an 'inus' condition of the occurrence of the fire: in other words, an insufficient but necessary part of a complex condition (short circuit, flammable materials, oxygen etc.) that was not necessary for the fire to occur but which on the given occasion was the only sufficient condition for its occurrence.[12]

---

[8] Heil in [42, p. 336] and Von Wright in [106, p. 114].

[9] Hitchcock in [45, p. 2].

[10] Hitchcock in [45, p. 2].

[11] Mackie in [65, p. 34] and in [66, p. 62]. Cf. also Sosa and Tooley in [89, pp. 8–9].

[12] Mackie in [65, pp. 33–34].

This definition of causation immediately solves the problem of imperfect regularities. Thus smoking, for example, does not always have to be followed by lung cancer in order for smoking to be the cause of lung cancer. Smoking is an 'inus' condition of lung cancer: hence, smoking is only invariably followed by lung cancer in certain conditions. However, the problem of spurious or coincidental regularities is left unresolved. Thus the light or the sound of an elevator, for example, could be an 'inus' condition, and hence the cause, of the elevator reaching its destination floor.[13]

Two further problems for this theory deserve brief attention. First, the theory is unable to account for the direction of causation. It seems as though causation has a direction in the sense that, if A is the cause of B, then B cannot be the cause of A: i.e., cause and effect seem to be asymmetrically related. However, if smoking, for example, is an 'inus' condition for lung cancer, then it seems to follow that lung cancer is also an 'inus' condition of smoking. To solve this problem by claiming that causes always precede their effects in time is to assume an analytical relationship between cause and effect which is contradicted by the logical possibility of backwards causation.[14] Second, the theory does not allow for the possibility of indeterministic causal relations in the sense of A being the cause of B without B invariably following on the occurrence of A.[15] Thus if A is an 'inus' condition of B, then B invariably follows on the occurrence of A. Since quantum mechanics has already established the reality of indeterministic processes – though only in the micro-world of atoms – the strength of this objection seems to depend mainly on the controversial issue of whether or not causation may be indeterministic. If this can be accepted, then the 'inus' theory must be abandoned. A possible alternative theory of causation would then be a probabilistic theory of causation, in which the central idea is that A is a cause of B if A raises the probability of B.[16] Leaving aside probability theories of causation, we will now move on to consider another approach to causation, known as manipulability theories.

---

[13] Mackie in [65, p. 39] and [66, p. 196ff.] acknowledges that accidental regularities pose a problem for his 'inus' theory.

[14] Note that backwards causation may be logically possible but contingently impossible. Mackie in [65, p. 50ff.] and [66, p. 160ff.]. Cf. also Hitchcock in [45, p. 3] and Faye in [24, pp. 20–46].

[15] Cf. the definition of determinedness provided in Section 5.3.

[16] Hitchcock in [45, p. 5ff.].

A fairly prominent manipulability theory is partly developed in response to the asymmetry of the causal relation referred to above.[17] The theory does not dissociate the notion of causation from the notion of regularity – and as such is not opposed to regularity theories – but rather takes as its starting-point the question of what confers on an observed regularity between events the character of a causal connection and thus distinguishes it from a mere accidental regularity. The answer is simply that in causation the regularity has a natural necessity to it. That is, suppose we observe some regularity between events p and q in the sense of q following immediately on the occurrence of p, then, according to the theory in question, the observed regularity is causal in nature if it holds of any occasion on which p did not occur that q would have followed, had p occurred. To put it slightly differently, for a regularity to have the character of a causal connection it must hold of a physically possible world, differing from the actual in p being the case, that if p is present, then q obtains.[18]

To understand this requirement of counterfactual regularity, it is necessary to have a model for the progression or development of the world. The model advanced within this theory depicts the world as at every moment in history being in a certain state with a number of causally possible futures. The possible futures are divided into one reflecting a natural, fully determined progression of the physical world in the absence of any human intervention and those representing the development of the world where there is human intervention. The model may be illustrated thus:

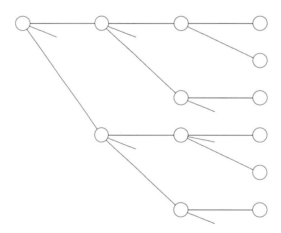

[17] Von Wright in [106, pp. 107 and 118]. Cf. also Von Wright in [107, p. 74ff.].
[18] Von Wright in [106, p. 114].

The nodes represent the total state of the world at a given point in time. The straight line following each node reflects the course of the world if it is left to its own natural, determined progression; hence the branching represents the course of the world if human intervention is made.[19] If the course of the world is tied to human intervention in the manner specified here, then the counterfactual requirement stated above may be reinterpreted. Thus the requirement that it must hold of a physically possible world, differing from the actual in p being the case, that if p is present then q obtains, is the equivalent of the requirement that humans on occasions on which the natural course of the world does not contain the occurrence of p, are able to bring about q by bringing about p. With this reinterpretation of the counterfactual requirement we end up with the following definition of causation: p is a cause relative to an effect q if, and only if, by doing or suppressing p 'at will' we could bring about or prevent q – or if, by producing changes in p 'at will' we could bring about changes in q.[20]

As mentioned above, this conception of causality readily solves the problem of accounting for the asymmetry of the causal relation. Thus the cause is simply distinguished from the effect on the grounds that we manipulate the former in order to influence the latter. Another advantage of this conception is the appealing nature of the implied epistemology of causation. Thus the process of coming to know that p is the cause of q is fairly straightforward – all it requires is experimentation. Suppose a state of the world in which p is not the case, and suppose that this is a state of which one can be fairly confident that if left to itself the world will not on some subsequent occasion bring about p. (Although fallible, this confidence is really the prerequisite of action. It is simply a conceptual feature – a counterfactual element – of every action that the changes it brings about would not have occurred unless the action was performed).[21] Given such a state of the world, we may go on to produce p and then see if the world, if left to its own development, brings about q. If so, this is not, however, sufficient to establish p as the cause of q. In order to rule out the possibility of the world by itself producing q on the relevant occasion, we may produce a state of the world in which p is not the case and then see if q follows immediately after or always follows on such occasions. Provided that this is not the case, we may thus be fairly confident that p is the cause of q.

---

[19] Von Wright in [106, p. 115]. Cf. also Øhrstrøm and Hasle in [73, p. 180ff.] for an elaboration of the notion of branching time.

[20] Von Wright in [107, pp. 70–74] and [106, p. 118].

[21] Von Wright in [106, p. 115].

It may be held against this theory, first, that it seems to allow for an event to be the cause of another event only if human agents are in fact able to bring about the one event by bringing about the other. However, there seem to exist causal relations among events which cannot be brought about or prevented by human action, such as the destruction of Pompeii by the eruption of Vesuvius or the gravitational pull of one galaxy on another.[22] One possible answer to this objection is that such events are complex in the sense of being composed by a number of other events, which may be causally connected in exactly the sense that we may manipulate the one by influencing the other; for example, in the case of the destruction of Pompeii by the eruption of Vesuvius it may be broken into other events such as the killing of a man by a stone falling on his head, a roof collapsing under a certain weight and so on.[23] Second, a common objection raised against manipulability theories is that the notion of bringing something about or of manipulation is in itself a concept that may only be explicated on the basis of the concept of causation. An action simply causes its outcome. Hence to define causation in terms of manipulation is simply circular.[24] In response to this accusation of circularity, it has been claimed that the outcome of an action is an essential part of the action – the outcome is intrinsic to the action and not extrinsic as is the case in causal relations.[25] Actions are simply *per se* a fundamental feature of the world.

Life-world

The concept of life-world is difficult to explicate fully. Nonetheless, we will here attempt to shed light on those aspects of the concept of life-world that have particular significance for the understanding and application of the hypothesis as stated above. This will involve the development of a concept of life-world different from other better-known concepts. At the end of this paragraph, however, we will briefly consider the differences between the concept of life-world developed here and these more well-known concepts.

---

[22] Von Wright in [107, p. 70]. Cf. also Woodward in [105, p. 4].

[23] Woodward in [105, p. 4] takes Von Wright to be saying that complex events are similar to other events, whereas the actual claim of Von Wright is that the complex events may be broken down into other events.

[24] Woodward in [105, pp. 2–3].

[25] Von Wright in [107, pp. 67–68].

Intuitively a life-world is something related to a particular individual's life – a world at the centre of which is the everyday life of a particular individual. This intuitive understanding captures at least two important aspects of the way in which 'life-world' is defined here. In the first place 'life-world' is taken to be subjective in the basic sense according to which something (judgement, experience, emotion etc.) is subjective if it is somehow dependent on the particular and contingent constitution of an individual, and because of that may vary from one subject to another.[26] Second, life-world is also defined in relation to the everyday life of the individual. That is, the life-world of an individual is something that relates to a basic predominant mode of life. With these remarks let us move to the definition of life-world.

To begin with, let us say that the life-world simply has to do with the world in which each particular agent unfolds his or her life. From an objective perspective the unfolding of life clearly has several dimensions – or, to put it slightly more picturesquely, several arenas at its disposal. These dimensions or arenas range from the external dimensions such as the physical environment over the social platform of family, local communities and global society to the internal dimensions such as the realms of thought, emotions, desires, will and so on. It seems as though the unfolding of life is constituted by series of consecutive events occurring in these (or other) dimensions. Thus the life of an agent may be viewed as a sequence of physical events such as birth, eating, sleeping, walking and so on, but also as a series of emotions such as anger, sadness, joy, indifference and the like, and a series of thoughts with a certain content such as 'the banana tastes better than other fruits', 'the girl next door is beautiful', 'the president is an idiot' and so on. This exemplification of a description of an agent's life in its physical, emotional and intellectual dimensions may obviously be extended to any possible dimension of the agent's life. As this concept of the dimensions of life serves only an auxiliary purpose in our attempts to define the concept of life-world, we will not take this investigation any further but merely note, once again, that it seems as though the unfolding of life may be described as series of events in these internal and external dimensions.

Turning to the task of more specifically defining the life-world, let us start by noting that not every event in any of the dimension of an agent's life is an event in an agent's life-world. Thus 'life-world' in the subjective sense it is to be given here concerns those events with a special relevance for a particular agent's life. More specifically, we will define 'life-world' as the 'world' of which it holds that, if an

---

[26] Derived from Brink in [10, p. 21].

event occurred, occurs or will occur in any of the external or internal dimensions of a particular agent's life to which the agent attributed, attributes or will attribute a significance making it particularly influential in shaping the everyday life and actions of the agent, then the event occurred in the life-world of the agent. Note that, defined in this way, an agent's life-world is an abstract entity. It is a collective noun for a certain class of events in the dimensions of an agent's life: namely, those that are considered by the agent to be strongly influential in shaping her everyday life and actions. Note also that, given the definition of life-world an event in the external dimensions, an earthquake, for example, must be followed by corresponding event in the internal dimensions with something of the same content, such as the thought that 'there is an earthquake going on', in order for that particular event in the external dimension to be part of an agent's life-world. Otherwise the agent will obviously not be aware of the event in the external dimensions and will therefore attribute no significance to the event. However, the opposite does not hold. Thus an event in the internal dimensions with a certain content, such as the thought that 'there is an earthquake going on', obviously does not have to be accompanied by an event in the external dimensions with something of the same content in order for the thought to be part of an agent's life-world.

One important element of this definition needs further elaboration. The definition trades both on the idea of agents or individuals attributing significance to events in the different dimensions of their life and on the idea of this attribution being formative for the agents' or individuals' lives and actions. Let us start by considering the notion of an agent attributing significance to an event. As used here, to attribute significance to an event is consciously to rank the event according to its (possible) influence on the fulfilling of one or more purposes rooted in desires, responsibilities, interests, values, principles, preferences, habits, emotional needs or the like that an agent may entertain either because they have freely chosen to, because of social pressure and manipulation, because of the laws of nature or perhaps because of a mixture of these.[27] In other words, an agent with the habit of smoking may attribute considerable significance to the event of getting onto a non-smoking flight, where this attribution is the equivalent of consciously considering the event to have a major influence on the possibility of fulfilling the purpose rooted in their habit of smoking: namely, to get a cigarette within a certain interval of time.

---

[27] In terms of the last option cf. the Kantian separation of the noumenal and the phenomenal, which allows for free will in the sense of actions originating in pure reason without compromising the principle that every event has a cause. Kant in [53, pp. 96–97].

Before addressing the presupposed ideas of the concept of life-world let us briefly examine a presupposition of this definition of attributing significance. As defined here, the act of attributing significance to an event is tied to the concept of purposive or intentional behaviour, in that the attribution of significance is allegedly something an agent does in the process of striving to fulfil a given purpose. This presupposition of the concept of attributing significance to an event fits very well with assumptions made throughout earlier chapters. Thus our model for the explanation of human behaviour is targeted at the explanation of human intentional or purposive behaviour. We will therefore go along with this assumption in acknowledging the possibility of human behaviour being purposive.[28] With the possibility of purposive behaviour we must now address the two remaining issues: namely, the agents' alleged ascription of significance to events and the alleged influence of this ascription on the agents' actions and lives.

Let us deal with these claims on the basis of a study of two cases supposedly reflecting real-life situations. Consider the case of an agent queuing in front of a railway ticket office in order to buy a train ticket as soon as possible and go home to her family. Suppose the agent realizes that for some time the queue of the adjacent ticket office has been moving faster because the clerk at that counter, unlike her colleague, is using both hands to serve the customers. At this point the agent will clearly attribute a certain significance to this event – the adjacent queue moving faster – in the sense of consciously ranking it on the basis of its possible influence on the fulfilling of a purpose: namely, the purpose of buying a ticket and going home as soon as possible. Moreover, the agent's attribution of significance will also shape her life and actions in the sense of giving rise to a number of deliberations and evaluations that may be followed by emotions such as annoyance and indignation and possibly also by actions such as changing queue and, in the longer term, avoiding that particular ticket office and passing on advice not to use that particular ticket office and so on. Consider another case, in which an agent aiming to live a joyful life discovers that the mental act of imagining a field of roses fills her with joy. Again, having made this discovery, the agent will not only attribute a certain significance to this event – namely, the event of the particular mental image of field of roses filling her with joy, in the sense of consciously ranking it on the basis of its possibly influence on the fulfilling of a purpose, i.e. the purpose of living a joyful life – but the agent's

---

[28] Aristotle in a famous passage from the *Nichomachean Ethics* states that 'Every skill and every inquiry, and similarly every action and rational choice, is thought to aim at some good; and so the good has been aptly described as that at which everything aims'. Cf. [3, 3].

attribution of significance will possibly also shape her life and actions in the sense of giving rise to considerations of the relationship between intellect and emotions, to increased focus on the nature of her thinking, to discussions with friends and perhaps even philosophers and so on.

If it can be granted that the above cases may, at least to some extent, reflect real-life situations, then we have some support for the claim that agents attribute significance to events in their internal and external dimensions, and that this attribution of significance may influence their life and actions. And if we do attribute significance to events, then, according to the definition, we have a life-world. Having thus developed a concept of life-word and given some reason to believe that it actually fits with our everyday practices, let us conclude with a brief comment on other conceptions of life-world.

The notion of life-world developed here differs considerably from other conceptions of life-world. In one such conception life-world is given a more social character, in being taken to denote the set of shared and implicit understandings, values and skills within a social group. The life-world is simply a shared resource and as such is constituted by a group.[29] It should be clear that this notion is not equivalent to the notion of life-world developed above, for the simple reason that the notion developed above clearly takes the life-world to be something the agent constitutes herself and after having accessed the basic conceptual understandings, values and skills of a social group. With these remarks in mind, let us move on to consider the claims inherent in the second hypothesis.

6.2.2 Belief, reality and causality

The second hypothesis makes three important claims. It claims that the strength of our belief in the reality of a particular patient with whom we are interacting is linked, first, to the belief that our actions will be the cause of an effect on the patient, and, second, to the belief that the effect of our actions on the patient will in turn also be formative for our own social life or life-world. The third claim is that both these beliefs are linked to our having some kind of evidence of their content.

As remarked in the outline of the claims of the first hypothesis, there seems to be no *prima facie* reason to suppose that the belief in having an effect on the patient and in this effect being formative for one's life-world is formed in a way that is different

---

[29] Habermas in [41, pp. 135–136].

from the formation of other beliefs about the patient. Hence the third claim may be taken to assert that we generally form beliefs about the patient on the basis of what we consider to be relevant and reliable evidence. The third claim is, therefore, shared between the first and third hypothesis, and will therefore be dealt with in the final section of this chapter.

The remainder of this section is dedicated to the analysis and discussion of the first and second claims. The two claims will in turn be analysed and discussed in two steps: first, the logical character of the relevant claim is explored; second, an attempt is made to support the relevant claim in and through the consideration of a fictitious case.

Believing and causal relations

The hypothesis states that any agent believes a particular patient to be real – is convinced of the reality of a particular patient – to the extent they come to believe that their actions are the cause of an effect on that particular patient. Before moving further into the discussion of this claim, it must be clarified on a number of issues.

First, we have to investigate the exact content of the belief in one's action being the cause of an effect on a patient. To put it in the form of a question: when agents believe their actions to be the cause of an effect, what is then believed? Our investigations of the notion of causation may be of help to us in answering this question. We will here assert that the belief in one's actions being a cause of an effect on a patient has a number of components, which may be revealed by means of the regularity and manipulability theories accounted for above. These components are taken to hold for a belief in causation formed in the course of everyday life and for a belief in a causal relationship between events on the macro-level.

We will deal first with the sufficiency or invariability component. It seems to be part of the content of the belief in one's actions being the cause of an effect that a given action simply is sufficient for bringing about a given effect. That is, if one believes one's action to be the cause of an effect on a patient, then it seems as though what is believed really is that from a certain action a certain effect invariably follows, and not that the action increases the probability of a certain effect. For instance, if we believe that our gentle kicking of a partner's leg under the table during a game of cards was part of what caused that partner to yell, then clearly what is believed is that the yelling invariably followed from the kicking etc. Second, there is the cause-as-condition component. It also seems to be part of the content of the belief in our actions being the cause of an effect that the cause is not the action in itself but rather the combination of the action and relevant conditions. For instance, having gently kicked the person's leg and heard the person yell, we will naturally believe the cause of the yelling to be a combination of the power of the kick, the point of impact, the pain threshold of the other person, the level of irritation owing to the course of the game and so on. A conditional analysis of causation simply seems to be part of believing something to be a cause of an effect. Third, there is the counterfactual component. It also seems to be part of the content of a belief in our actions being the cause of an effect that the effect would not have occurred on that particular occasion if the action had not been performed. If the effect had been achieved without the action being performed, then there would have been some other set of conditions sufficient for the effect to occur; hence it cannot be that the action performed in the given circumstances was the cause of the effect. In other words, to believe that the gentle kicking of the partner was in part the cause of her yelling clearly involves the belief that the partner would not suddenly have yelled on the given occasion if she had not been kicked on her leg. Fourth and finally, there is the manipulability component. It seems to be a very ingrained belief that the causal relation is fundamentally such that the effect may be influenced by manipulating the cause – especially in cases where the cause involves human actions. This means that to believe that the gentle kicking of our partner was in part the cause of her yelling clearly involves the belief that by manipulating the cause, namely the power and aim of the kick, the agent would influence the effect, so that it would amount to no more than a wild look and no yelling. Note that the third and fourth components are related in the case of action. Thus in the event that one holds the belief that a certain effect would not have obtained if one had not acted in a certain way, then it seems as though one may infer that the effect is manipulable, in the sense that one may bring about the effect and change the effect by one's actions.

The four components outlined here may obviously be more or less explicit in the formation of the belief that one's action on the given occasion was the cause of an effect on the patient. Furthermore, it may also be the case that certain other components are of relevance for the formation of the relevant belief. However, we will here and in what follows take these components to be a prerequisite of forming the belief in one's actions being the cause of an effect; in other words, should agents fail to believe any of these components to be satisfied or come to believe in the negation of any of the components, then they would not believe their action to be the cause of a relevant effect on the patient.

Having clarified what seems to be involved in a belief of being causally effective, we may now turn to a clarification of the overall nature of the first claim of the second hypothesis: namely, the claim that an agent believes a particular patient to be real to the extent that he comes to believe that his actions are the cause of an effect on the patient. It is worth noting that this claim is intended in a stronger sense than the first claim of the first hypothesis. Thus we will assert that the belief in causally affecting a patient is unnecessary but sufficient for an agent to believe – or to be convinced of – the particular patient being real. This implies that it is fully consistent with the first central claim of the second hypothesis that an agent comes to believe in the reality of the particular patient with whom the agent is interacting without believing his actions to be causally affecting the patient, but not that an agent may believe himself to be causally affecting the particular patient without believing the patient to be real. It is simply an ingrained belief that we cannot have an effect on something particular that does not exist. Hence if we believe we have had an effect on something particular, then it inevitably follows that the reality of this particular entity will not be questioned by the normal agent engaged in everyday dealings. Bear in mind that we are here making a claim regarding an agent who is not being deceived by a mad scientist or a hypnotist or making epistemological mistakes etc.

This interpretation of the first claim of the second hypothesis allows for an agent to believe in the reality of a particular patient on the basis of, say, having received a 'Mayday' message from the patient or being told about the patient by other agents considered to be reliable sources of information. That is, an agent may come to believe in the reality of the particular patient on the basis of being causally affected by the patient or on the basis of other people causally affecting or being affected by the patient – although one could perhaps argue that also in these cases it is really the subsequent belief that one may causally affect the patient that gives rise to the belief of the particular patient being real. On the other hand, however, the interpretation of the first claim given above does not allow for an agent to believe he is causally affecting a particular patient without believing the patient to be real. Hence at the core of the second claim lies the assumption that, whatever else an agent may believe about a patient, it is not possible for him to believe that he is causally affecting the patient and at the same believe the particular patient not to be real. As will later become clear, this also entails that an agent cannot believe that he may causally affect the patient without believing the patient to be real.

Before entering into the discussion of the first claim of the second hypothesis, it is worth stressing that it distinguishes between an agent *believing* that he is causally affecting a patient and an agent *causally affecting* a patient. It is clear that the claim concerns the former. It seems to be a fact that at any point in time any human being exerts a causal influence on any other physical entity in the universe simply because of the gravitational forces among masses. It is evident that it would be a mistake to assume that an agent causally related to any entity in the world believes any of these entities to be real. For this reason the claim only concerns those patients or entities whom or which an agent believes that he is causally affecting. By extension, the claim is also intended to distinguish between, on the one hand, an agent being a passive or active part of a causal relation and, on the other, an agent being part of a causal relation intentionally or unintentionally. Thus an agent believing that he is causally affecting a patient is here assumed to be an agent who is *intentionally acting* in order to influence the patient, where action, as defined earlier, involves bodily movement. Consequently the claim does not cover cases such as that above, in which an agent is part of a causal relation by virtue of mere existence as a physical object.

The mirage argument

The main argument advanced in this paragraph relies on the concept of a mirage. Before looking at this argument, however, let us briefly consider the *prima facie* credibility of the first claim of the second hypothesis.

On the face of it, the claim seems quite reasonable. It seems as though we may indeed in many of our daily interactions believe a particular entity with whom or which we are interacting to be real on the basis of believing we are causally affecting that particular entity. I may, for instance, believe the particular cup of coffee placed next to me to be real simply on the basis of believing that I have now for some time been affecting the state of the cup of coffee by the actions of picking up the cup and pouring the contents down my throat. Likewise I may believe the particular woman serving me the cup of coffee to be real only on the basis of believing that my complex action of ordering coffee affected the state she was in while standing behind the counter and smiling. Nevertheless, while these examples may lend some credibility to the claim, they do not really take us far. It may thus be objected that the belief in the reality of a particular entity is really a prerequisite of believing in causally affecting the patient. Applying this to the first of the examples provided above, the point would be that I come to pick up the cup and pour the coffee down my throat because I already believe the particular cup and coffee to be real; in other words, I believe the particular cup and the coffee to be real and therefore I believe I am able to affect them causally.

At first sight the cogency of this objection may seem limited. The gist of the argument is that the belief in the reality of a particular entity is a necessary prerequisite of attempting to causally affect the entity. This is a plausible claim. However, it is not very much to the point. The first claim of the hypothesis is that the belief in causally affecting the particular patient is sufficient for the belief in the patient being real – or, as the previous paragraph puts it, that it is not possible to believe in causally affecting a particular patient without believing the patient to be real. If we look at the logic of this claim, it is not at all inconsistent with the claim inherent in the objection. Thus if it holds that an agent must believe in the reality of a particular entity before attempting to causally influence the entity, then it is, other things being equal, also the case that we cannot believe we are causally affecting a particular entity without believing the entity to be real (there is no reason why as agents we should change our belief in the reality of the particular patient when moving from the belief that we *may* causally affect the patient to the belief that we *are* causally affecting the patient). Hence the objection clearly misses the mark, logically speaking.

Although the objection fails from a purely logical perspective, it offers a most welcome opportunity to develop further the first claim of the hypothesis. As we have just shown in the dismissal of the objection, the first claim is intended to incorporate the possibility of the belief in the reality of the particular patient being a necessary prerequisite of the belief in being able to causally affect the patient. However, if all there is to the role of the belief in causally affecting a patient is that this belief presupposes the belief in being able to have a causal influence on the patient, which again presupposes a belief in the reality of the particular patient, then it seems rather misleading to state the claim in the way we have done. Thus in stating that the belief in causally affecting the patient is sufficient for the belief in the reality of the particular patient we seem to be suggesting that the belief in the reality of the particular patient somehow *originates* in the belief in causally affecting the patient – or, at least, that the belief in causally affecting the patient *adds* something to the belief in the reality of the particular patient. The privileged role ascribed to the belief in causally affecting the patient is no mere confusion. We will argue here, therefore, that the belief in the reality of the particular patient presupposed in the attempt to causally affect the patient is not at all equivalent to the belief in the reality of the particular patient resulting from the belief in causally affecting the patient – at least, not when taking into account how these beliefs may be formed in situations of a particular kind.

It is at exactly this stage that the concept of a mirage becomes relevant. We will use the concept of a mirage in a small thought experiment, with the purpose of showing that the process of coming to believe in the reality of the particular patient on the basis of believing in causally affecting the patient is not a process in which evidence of the reality of the particular patient is simply added, but a process in which a very significant kind of evidence is added. In this way, then, the belief in causally affecting the patient comes to play a privileged role for the formation of our belief in the reality of the particular patient – at least, in situations of a particular nature. As a reminder it should be borne in mind that our concern throughout this book is the agents' deliberations and reflections on their beliefs: that is, agents striving to form justified beliefs.

Imagine a former Olympic competitor in the discipline of javelin-throwing having been robbed of his scarce supplies while crossing the Saharan desert and left to stagger his way through the burning desert. At some point during the day the man sees what seems to be an oasis in the distance and heads for it. Suppose now that, as he staggers towards the oasis, he discovers what seems to be another oasis only marginally further away but in a slightly different direction. The newly discovered oasis is positioned at the end of a gap between hills of sand and consequently allows for only a very limited view of it. Struck by doubt as to whether there are any oases anywhere and at the same time knowing that his strength is giving out and will suffice to bring him to only one of these possible sources of life, the javelin-thrower is overcome with despair. However, being a former Olympic competitor in his chosen discipline, he takes off one of his worn shoes, gathers all his strength and throws it towards the newly discovered oasis. As the shoe comes to the end of its trajectory, it seems to break the surface of water and cause a splash. The javelin-thrower then turns so as to the face the original destination and throws his other shoe towards it. Although the shoe seems to reach its target, there seems to be no visible effect.

Before extracting the lesson we would like to be learned from this case, a few comments may be helpful in understanding the interpretation of the case suggested below. First, the setting of a desert is chosen since it would seem natural to question the reality of the particular content of perceptions when subjected to the merciless sun for a prolonged period of time. Second, as the javelin-thrower at first discovers an oasis and sets out towards it, it seems he clearly believes in its existence. This belief in the reality of the particular oasis is in part founded on evidence of the oasis gained through mere observation: namely, the evidence acquired through non-causal relation with the oasis (cf. the previous paragraph on the notion of a causal relation to an entity). Third, the discovery of the second oasis and the combination of the shoe causing a splash in the one case and not in the other induces a belief in having causally affected the one oasis and not the other.

The argumentative use we will make here of the case of the javelin-thrower and the oasis rests on what seems to be three plausible suppositions. First, it seems plausible to suppose that the javelin-thrower, faced with the relevant evidence of the existence of these particular oases, would choose to head for the one where his shoe had appeared to cause an effect. Second, it seems plausible to suppose that the javelin-thrower's decision to head for the one oasis rather than the other reflects a difference in his beliefs regarding the reality of the particular oases. After all, given the nature of the situation, it is clearly the reality of the particular oases the javelin-thrower is questioning and deliberating about. Hence the decision to head for the one oasis is clearly the result of his conviction that this particular oasis is real and the other a mirage. Third, it seems plausible to suppose that the difference in the agent's beliefs regarding the reality of the particular oases reflects a difference in the agent's beliefs regarding his ability to affect the oases causally. Thus it seems plausible to suppose that the agent believes the one particular oasis to be real and the other unreal on the basis of his belief in having causally affected the one but not the other. Moreover, it seems as though the belief in causally affecting one of the oases and not the other is sufficient to convince the agent of the reality of the one particular oasis and not the other in the relevant circumstance.[30] Furthermore, it seems as though the difference in the mere observational evidence of the properties of the oases when observed from a distance is, so to speak, 'outweighed' by the evidence of causally affecting the one oasis and not the other. Thus the belief in the reality of the one particular oasis comes to be established, it seems, on the grounds of the belief in causally affecting the one oasis and not the other.

Given the plausibility of the three suppositions made above, the mirage example has fed into the discussion of the first claim of the hypothesis in two important ways. First, in terms of accounting for the difference between the belief in the reality of a particular entity presupposed in engaging with the entity (the cup of coffee) and the

---

[30] Cf. Dancy's distinction between sufficient reasons and circumstances in [16, p. 24].

belief in the reality of a particular entity based on the belief of having causally affected the entity (the oasis), we may conclude that in situations comparable to the mirage example the difference between these beliefs is really a difference in the evidence of the reality of the entity. Moreover, the difference in the evidence of the entity is not a straightforward difference in the amount of evidence available but a difference in the beliefs this evidence may assemble in support. Thus the evidence of causally affecting an entity is clearly not just evidence of the properties of the entity when it comes to forming beliefs regarding the reality of the particular entity – as such, it would add very little to the evidence already available, and anyway, the total amount of evidence of the properties of the original oasis would seemingly be greater. Second, in terms of grounding the first claim of the second hypothesis, these considerations have taken us some but not all of the way. The first claim is that the belief in causally affecting the patient is sufficient for the belief in the reality of the particular patient. The mirage example seems to show that in a certain kind of situation the belief in causally affecting an entity is sufficient for the belief in the reality of a particular entity. Moreover, it shows that the amount of evidence of the properties of an entity takes on less importance when there is evidence of having causally affected the relevant entity. It is exactly this feature of the example that may lend some credibility to the first claim of the second hypothesis. Thus if evidence of having causally affected an entity may render another kind of evidence – i.e., mere observational evidence, of less importance in coming to believe in the reality of a particular entity – one may suppose that the same would hold for yet other kinds of evidence relevant for the formation of the belief in the reality of a particular entity. One such kind of evidence could be the testimony of other people. Suppose the javelin-thrower is accompanied by a very good friend who tells him that he has actually been in exactly this place before and that he knows the oasis behind the hills of sand to be a mere mirage. The testimony of the friend is obviously relevant for the javelin-thrower's deliberations regarding the reality of the particular oasis. However, given that the javelin-thrower has evidence of having causally affected the oasis, it may still be plausible to suppose that he would believe the particular oasis to be real solely on this ground – and contrary to the testimonial evidence. This argument could possibly be further extended to other kinds of evidence.

Anyway, in the light of these considerations, the only warranted conclusion is that in certain kinds of situation – those similar to the one in the mirage example – the belief in having causally affected an entity is sufficient for the formation of a belief in the reality of the particular, causally affected entity. Hence, although we have advanced some arguments in support of the general validity of the first claim of the second hypothesis, the claim has not been shown to hold generally. A natural consequence would be to weaken the claim along the lines of the first claim of the first hypothesis. The claim would then be that causally affecting an entity plays an important – though neither a necessary nor a sufficient – role in the formation of a belief in the reality of the relevant entity. This weakened claim seems to have been strongly vindicated here.

Before moving on to consider the second claim of the first hypothesis, let us briefly try to put our reflections into a more general perspective. Thus the first claim of the second hypothesis may be seen as an attempt to capture the common intuition that in actively engaging in the world our conviction concerning the reality of the particular entities in it becomes stronger than a similar conviction resulting from passive experience. To the extent that this common intuition is sound, it obviously lends credibility to the first claim of the second hypothesis.

6.2.3 Belief, reality and life-world

As already outlined, this subsection is split into two parts. In the first of these the content and logical character of the second claim of the hypothesis are explored. In the second part an attempt is made to support the hypothesis by considering a case of interaction.

Believing in life-world effects

The question to be answered here is this: when any agent believes the effects of their actions on the patient to be constitutive of their life-world, what is then believed? Given the character of the life-world described above, we can now give a more exact interpretation of the relevant belief.

In the section alluded to we defined 'life-world' as a set of events in the past, present or future of which it holds that to each event is attributed a significance by an agent making it particularly influential in shaping the agent's everyday life and actions, where the attribution of significance was defined as consciously ranking events according to their (actual or possible) influence on the fulfilment of a given purpose. This means that an agent believing that the effects of her actions on the patient are constitutive of her life-world is the equivalent of an agent believing that she will consider the effects of her actions on the patient to be significant to a degree that will make them particularly influential in shaping her life and actions. Moreover, believing the effects constitutive of her life-world is the equivalent of believing the effects of her actions on the patient to be of such importance for her ability to fulfil one or more of her purposes in life that they will be influential in shaping her life and actions. Fitting this explication of belief in the effects of actions being constitutive of the life-world into the wording of the second claim of the hypothesis, we end up with the following statement: an agent believes the particular patient to be real to the extent that she comes to believe the effects of her actions on the patient to be of such importance for her ability to fulfil one or more of her purposes that these will be influential in shaping her course of life and action.

In the section on the concept of life-world we went into some detail in our explorations. However, the present elaboration of the second claim in the light of the concept of life-world deserves a little attention. It is worth remarking that in the formulation that has just been given the second claim of the hypothesis opens up the possibility of the belief in the reality of the particular patient being dependent on a number of purposes, for whose achievement the effects of an agent's actions on the patient may be held to be important.

The point that several of an agent's purposes may be affected by certain events can readily be illustrated by revisiting the case of the agent queuing in front of the railway ticket office in the earlier analysis of the life-world. As the example was then developed, the only purpose affected by the idle clerk in the ticket office was the agent's purpose of buying a train ticket and go home to her family as soon as possible. Suppose now that the agent queuing also entertained the purpose of exhibiting patience and forbearance when faced with what may be described as the imperfections of other people. In this slightly more complex situation it is clearly the case that a greater number of the agent's purposes are affected by the event of the employee serving the customers using only one hand. In accordance with our original exposition of the example, the agent would in this case attribute significance to the event on the basis of the two relevant purposes. First, she would rank it on the basis of its (possible) influence on the purpose of returning to the family as soon as possible, and second, she would rank it on the basis of its (possible) influence on the general purpose of exhibiting patience and forbearance. Finally she would probably rank the situation as a whole in terms of its conduciveness to the fulfilment of the set of desires relevant in the given situation, and this ranking of the situation as a whole would then be the one to influence her course of life and action.

In order to avoid any confusion, the point to be made by this extended version of the example introduced earlier is simply that the occurrence of a certain event may be important for the fulfilment of more than one purpose. Thus, returning to the second claim of the hypothesis, the effects of an agent's actions on the patient may be rendered significant by the agent for the fulfilment of more than one purpose. Hence, if the belief in the reality of the particular patient is dependent on the significance attributed to the effects of the agent's actions on the patient, then it comes to be dependent on the significance of these effects for the fulfilment of each of the relevant purposes.

In terms of the overall logical structure of the second claim, it is worth noting that it clearly incorporates the first claim of the second hypothesis. As it stands, it makes the belief in the reality of a particular patient dependent on a particular effect of an agent's actions on the patient, namely an effect that would be constitutive of the agent's life-world. Thus, according to the second claim, the belief in the reality of a particular patient is tied to the agent's actions having an effect on the patient – just as the first claim states. In the section on the logical structure of the first claim we argued that the belief in having an effect on a patient is not necessary but sufficient for the formation of the belief in the reality of the particular patient. Hence it follows that an agent's belief in her actions having an effect on the patient that is constitutive of the agent's life-world will at least be sufficient for the belief in the reality of the particular patient.

In opposition to the arguments just presented, it may perhaps be tempting to argue that it is necessary for the effects to be constitutive of the life-world of an agent for the formation of the belief in being the cause of an effect on the patient at all. The point would be, then, that it is exactly because of the occurrence of an event that agents consider to be of consequence for their life-world that they come to realize that they are actually having an effect on the patient. This position may be refuted by the following example. Imagine a young man visiting his friend for a game of cards. The friend is most preoccupied with the planet Mars and has actually, by means of a sophisticated set-up of computers, managed to take control of a rover left by NASA on the surface of the planet. Suppose now that the space-geek encourages his visiting friend to manipulate the rover in between their games of cards. Being completely uninterested in anything remote from the earth, the friend hesitatingly accepts the invitation, provided he can return straight afterwards to the card table. They then go to the computers, the visiting friend uninterestedly manipulates the rover by means of the arrow keys on a keyboard and they return to the card table. The point of this example is that it seems plausible to suppose that the agent, uninterestedly manipulating the rover on the surface of Mars, actually formed the belief that he was the cause of an effect on the rover without, however, that effect on the rover being at all constitutive of his life-world.

This interpretation may be contested on several grounds. For one thing, it is not clear that his effect on the rover was not constitutive of his life-world. This objection presents a welcome opportunity to elaborate on the notion of life-world as well as to emphasize an important feature of it. The life-world is defined as a set of events of which it holds that each event is attributed significance by an agent, making them particularly influential in shaping the agent's everyday life and actions. Before dealing with the objection, it is worth noting that the definition of life-world encompasses two possible meanings of an event being constitutive of the life-world. One the one hand, the occurrence of an event may be constitutive of the life-world simply because of its occurrence. On the other hand, the occurrence of an event may be constitutive of the life-world because of the content of the event. So far we have more or less exclusively applied the concept of life-world using the latter interpretation of an event being constitutive of the life-world. Applying the distinction to the example just given, the visiting friend may believe that the event of manipulating the rover is constitutive of his life-world simply because of manipulating the rover or because of the content of these manipulations of the rover: e.g., turning right or left.

Returning to the objection made above, it must be rejected for the following reason. Although it may be argued that the visiting friend may assign a significance to the occurrence of the event or to the content of the event of manipulating the rover on Mars in terms of its influence on his ability to fulfil certain of his purposes – playing cards, for example – it is not at all clear that he would consider the event particularly influential in shaping his course of life and actions – persuading him, for example, to decide never to do anything like it again in the middle of a game of cards or perhaps returning to the table in an angry mood. It seems once again plausible to suppose that he may consider the event fairly insignificant for his purposes in the sense of leaving the event without it shaping his course of life or action in any of the ways outlined here. The more general point to be made here is that, although we are, as humans, seemingly always engaged in the pursuit of the fulfilment of certain goals and purposes and although the occurrence of any event will therefore affect this pursuit in one way or another, this does not entail the relevant event being an event in our life-world as defined here.

Having thus refuted the position according to which the belief of the effects being constitutive of the life-world of an agent is necessary for the formation of the belief in being at all the cause of an effect on the patient, let us briefly comment on the possibility that the occurrence of an event that agents consider to be of consequence for their life-world is sufficient for them to be aware of actually having an effect on the patient. As is fairly obvious, such an event cannot be sufficient to create awareness of being the cause of an effect on the patient since there are events occurring which agents may consider to impact on their life-world but which do not involve the agents being the cause of an effect on a patient.

Before ending these reflections on the nature of the overall logical relationship between the agents' belief in the reality of a particular patient and the agents' belief in the effects of their actions on the patient being constitutive of their life-world, it remains to be settled what difference it makes for the belief in the reality of the particular patient that agents believe not only that they are having an effect on the patient but also that this effect is constitutive of their life-world. The simple but very important answer is this: the agents' belief that the effect of their actions on the patient is constitutive of the agents' life-world adds to their conviction that the particular patient is real: i.e., it adds to the certainty with which they hold the belief that the particular patient is real.

The stockbroker argument

The main argument to be advanced in this paragraph uses as its backdrop the scenario of a stockbroker investing on the stock market. The argument to be developed throughout this section is noticeably different from the arguments put forward previously. Those arguments all related an agent's belief in the reality of a particular patient to the agent coming to believe in certain objective ontological conditions obtaining – either the determinacy of the patient or a causal effect of the agent's actions on the patient. The argument to be presented here relates an agent's belief in the reality of a particular patient to the subjective significance an agent may attach to the fact of affecting a patient in a certain way.

Imagine a stockbroker having faithfully followed his trade everyday of his working life. Everyday he has been coming into his office and skilfully investing vast amounts of money on behalf of a variety of clients. Suppose now that, as the stockbroker enters his office on the day of his retirement, he is called for by his boss. The boss proudly announces that on this his very last of work the stockbroker will be given a vast amount of money to invest, as a sign of appreciation for his loyal and competent work throughout his career, and that he will be allowed to keep the surplus generated at the end of the day for his retirement. Suppose also that the stockbroker has for some time planned a number of goals to pursue after his retirement, and that a prerequisite of the achievement of these goals is a certain level of funds, towards which the surplus of a day's work will make a decisive contribution. More specifically, let us suppose that the stockbroker has planned, among others things, to take his grandchildren to the zoo more often and to start collecting diamonds after his retirement.

Before making explicit what we would like to extract from this example, a few comments may be helpful for the interpretation of the example underlying this illustration. First, again, the example of the stockbroker investing money has been chosen since the question of the reality of the money being exchanged on the stock market seems a natural one to pose. Second, it seems as though a stockbroker investing money on behalf of individuals and companies must and clearly will believe in the existence of money. Taking into account our considerations about the first claim of the second hypothesis, this belief in the reality of money may be founded in part on evidence of causally affecting representations of the money – e.g., numbers on a screen – and through evidence of affecting individuals and companies through affecting the representations of money, but it may also be founded in part on the mere observation of money or some other relevant features. Third, it is clearly the case that the event of being given a vast amount of money to invest for his own benefit on the last day of work is an event in the stockbroker's life-world. In the light of the stockbroker's goals for the time after his retirement, therefore, he will clearly attribute significance to the event of generating a surplus on the basis of the event's possible influence on his achievement of the relevant goals, and this makes it particularly influential in shaping his course of life and actions.

The case of the stockbroker may come to count in favour of the second claim of the hypothesis on the basis of three suppositions all enjoying a certain *prima facie* plausibility. First, it seems plausible to suppose that the stockbroker would relate differently to the money he is given on the last day at work from the way he has been relating to the money he has invested throughout his career. Second, it seems plausible to suppose that the difference in the stockbroker's relation to these 'two kinds of money' may – for all practical purposes – be described as a difference in his conviction concerning the reality of the particular money. Thus it seems plausible to suppose that he may have been less convinced of the reality of the particular money he has been investing prior to the last day at work than of that invested on the last day at work. Let us briefly elaborate. In both cases the stockbroker will have attributed certain properties to the money with which he has been dealing. One such property could very well be the property of 'buying power': that is, money has the property of endowing its owner with buying power to an extent determined by certain other factors. The claim made here is simply that the stockbroker will be more certain of the reality of the money as something possessing among others the property of 'buying power' when it comes to the money invested on the last day than the money invested prior to that day. Third and finally, it seems plausible to suppose that the difference in the convictions concerning the reality of the particular money is related to a difference in the influence that the 'two kinds of money' exert on the life of the stockbroker, and hence that it reflects a difference in their relationship to the life-world of the stockbroker. Thus it seems a plausible account of the difference in beliefs regarding the reality of the particular money that it is brought about by the fact of the money invested prior to the last day of work not having a clear and distinct effect on the stockbroker's pursuit of certain goals considered to be significant to a point of making them particularly influential in shaping his life and actions. On the contrary, on the last day at work every single penny of the money invested will have a clear and distinct effect on the stockbroker's pursuit of certain goals, which the stockbroker may certainly consider to be significant to a point that makes them particularly influential in shaping his life and actions.

Given the plausibility of the three suppositions made above, the case of the stock-broker serves its purpose. The purpose of this section was to ground the claim that the agents' belief that the effect of their action on the patient constitutes their life-world adds to their conviction that the particular patient is real. The case of the stockbro-ker developed above has clearly taken us some of the way towards showing this. The reason for the reservation regarding the strength of the stockbroker example is based on the objections raised below.

The first objection concerns the interpretation of the example. In the interpretation it seems to be assumed that the daily investment of money prior to the last day at work is not at all constitutive of the stockbroker's life-world. If the daily job of investing money were granted to be constitutive of the stockbroker's life-world, there seems to be no reason why the effect on the life-world should provide the stockbroker with reason to have different beliefs concerning the reality of the 'two kinds of money'. The answer to this objection is simply to claim a difference in degree. Thus although the money invested prior to the last of work may influence the stockbroker's other goals by being part of the job earning him a salary, there is, other things being equal, no clear, distinct and correlated effect of the particular investments made at work on his salary. Hence, it seems as though the stockbroker will be more convinced of the reality of the particular money invested on the last day at work, since this investment will have a clear and distinct effect on the possible pursuit of certain goals. More generally, the answer to the advanced objection is that, although all events may be measured against an agent's purposes in life,[31] the mere possibility of this measurement does not render a difference in the degree of influence on the agent's purposes impossible or unlikely. Let us now turn to the second objection.

Intuitively the stockbroker example seems to support the point that a stockbroker may relate differently to the money he is investing under varying conditions, and that this difference in his relation may be caused by a difference in the degree to which the two events of investing money may be constitutive of his life-world. Although it appears a fairly plausible interpretation of the example, it is perhaps questionable whether the difference in the relation to the money invested on the two relevant occasions is equivalent to a difference in his convictions concerning the reality of the particular money. The question really is whether the belief in the reality of the particular money plays a key role in the stockbroker's different ways of relating to the money. Perhaps the example is better accounted for using other concepts, such as relevance, vividness, closeness or the like. That is, perhaps the difference in the stockbroker's different ways of relating to 'the two kinds of money' has to do with the stockbroker taking the money invested throughout his career to be somehow more 'distant', 'irrelevant' or 'less vivid', whereas he may believe the money invested on the last day of work to be somehow 'closer', 'relevant' or 'vivid'. The answer to this objection is that the suggested alternative accounts of the difference in the stockbroker's ways of relating to 'the two kinds of money' do not exclude the one provided here. Thus it may very well be that the stockbroker is more convinced of the reality of the particular money invested on the last day at work precisely because it seems 'closer', 'more relevant' or 'more vivid'. Remember again that in the previous chapter we construed the notion of believing in the reality of a particular entity as that of believing an entity to possess a certain set of properties. In the case of the stockbroker, he may believe money to be characterized by the property of 'buying power', among others. The point is that when the stockbroker comes to be more

---

[31] Cf. the distinction between the occurrence of an event and the content of an event introduced above.

convinced of the reality of the money invested on the last day of work – that is, convinced that it possesses 'buying power' – then this may very well have to do with this money appearing 'closer', 'more relevant' or 'more vivid' – where the closeness, relevance and vividness then would in turn be explained with reference to the influence the money invested exerts on the life-world of the stockbroker.

6.3   Hypothesis III: Reality and vulnerability

The preceding sections have seen the investigation of the first and second of the hypotheses introduced in the previous chapter. We now turn to the third hypothesis. In summary, this claims that our sense of the reality of a particular patient of our actions is linked to a belief in the vulnerability of the patient.

The investigation basically has two parts, and these are reflected in the overall structure of this section. First there is an analysis of the notions of vulnerability and interdependence; this is followed by an analysis and discussion of the asserted relation between our sense of the reality of a particular patient and the belief in the vulnerability of the patient.

6.3.1   Vulnerability and dependency

For the sake of clarity of focus let us start by restating the third hypothesis and apply it to the special case of the patient being another human being. The third of the hypotheses states that:

[3. Hyp.]   An agent engaged in interaction with a patient is convinced of the reality of the particular patient of her actions to the extent that she, on the basis of what she in the given situation considers to be relevant and reliable epistemological evidence, comes to believe that the patient is vulnerable to or dependent on her actions.

In the special case of the patient being another human being or person the hypothesis simply claims, to put it rather simplistically, that our belief in the reality of the person with whom we are interacting is constituted by the person appearing to us as vulnerable and dependent on our actions: e.g., that the person is or may be harmed by certain words or deeds or that the person may be in need of something we may provide through our actions.

Vulnerability, the good and practical reasoning

The notions of vulnerability and dependency play a crucial role in the hypothesis as worded above. In this section we explore the concepts of vulnerability and dependency in order to gain a clear understanding of what is meant here by an agent coming to believe in the vulnerability and dependency of the patient. As with the concepts of determinateness, causation and life-world investigated in earlier sections, we are not hereby assuming that this belief reflects a conscious awareness of theoretically constructed concepts of vulnerability or dependency, but rather that these concepts may be helpful in simply circumscribing a state of affairs that may to a greater or lesser extent be believed to have obtained when an agent believes a patient to be vulnerable or dependent.

As a first observation, it should be noted that the vulnerability of any kind of entity basically means that the entity may be subject to harm or damage. As a second observation, it should be noted that, defined in this way, the vulnerability of, for instance, humans and animals is closely related to a notion of 'the good' for these entities or to what it means for these entities to flourish. Suffering harm or damage may thus be reconstructed as being the result of a hindrance or simply interference to the achievement of certain benefits that would otherwise be conducive to the flourishing of an entity.[32] This conceptual tie between human vulnerability and human flourishing poses a problem for our endeavour in this section, which is to circumscribe the state of affairs we consider to exist when an agent believes a patient to be vulnerable. Our strategy in the sections on the first and second hypotheses was to develop a theoretical view of key concepts such as determinateness, causation and life-world and then to claim that an agent who believes in, for example, causally affecting a patient is believing in the occurrence of the state of affairs described by means of the theoretical concept of causation. This approach rested on the seemingly reasonable assumption that we all more or less share a common concept of what it means for us to cause something. This strategy cannot be pursued here, for to ascribe to agents a more substantial belief in the vulnerability or dependency of a patient is to ascribe to the agents substantial normative beliefs regarding the human good, which may clearly vary from agent to agent. In other words, to ascribe to agents the belief that a patient is dependent on friendship is obviously to ascribe to the agents the belief that friendship is valuable or good – a belief that may vary from agent to agent. Hence if we are to ascribe to agents anything more substantial than the belief that the patient may be harmed by something taken by the agent to be good, then we are in need of some theory of human vulnerability and dependency that does not make substantial claims regarding the good of mankind.

In what follows we explore an account of human vulnerability and dependency that seems to satisfy exactly this requirement on the basis of reflections on what we

---

[32] MacIntyre in [63, pp. 63–64].

might call the basic human condition. This may in the end help to shed some light on the possible content of the belief of someone who believes the patient to be vulnerable without at the same time ascribing to this person a belief in some designated good that the person may not hold. It has to be added, however, that the account to be investigated below is also included because we take it to present an adequate account of certain basic human conditions.

As noted above, the concept of flourishing is closely related to the concept of 'the good'. Moreover, it is closely related to a particular use of the concept of 'the good'. Thus something can be good for someone by virtue of being a means to something further that is itself good. For instance, certain skills may be good as a means to the achievement of promotion, or to be at a certain place at a certain time may be good as a means to the achievement of something favourable. However, something may also be good to be or to have or to do for someone by fulfilling a certain role or function in a socially established practice. These benefits are internal to the relevant practices and are worth pursuing for their own sake, given the fact of participation in the practice. Strategic skills, for example, are good for someone who is a chess-player.[33] Although these are undoubtedly of benefit within a certain practice, the goodness of the practice may obviously be questioned. Doubt may be raised as to whether or not it is at all good to play chess. Consequently there is a third kind of good: something may be good to be or to have or to do unconditionally. That is, something may be good in its own right for someone *qua* human being: i.e., a good to be pursued for its own sake, regardless of a certain social role or function. Those forms of the good whose achievement are constitutive of human flourishing are of this kind. Thus human flourishing is obviously flourishing as a human and not *qua* a certain role or function.[34]

Having learned the character of those forms of the good whose achievement is conducive to human flourishing, the obvious question to ask is what is required to

---

[33] MacIntyre's concept of internal and external goods in relation to practices as a special kind of cooperative activity is developed in depth in [64, pp. 187–189].

[34] See MacIntyre in [63, pp. 66–68] for the three uses of 'the good'. MacIntyre also mentions a fourth kind of good, namely the satisfaction of bodily desires, which are taken to be a good only in infancy. Cf. [63, p. 68]. This distinction between a good conducive to human flourishing and a good constituted by the satisfaction of bodily desires seems to imply a denial of a Humean account of normative reasons investigated in a previous chapter, but also MacIntyre's rejection in [63, pp. 86–87] of Williams' account of normative reasons in [101, pp. 101–113]. However, MacIntyre accepts a Humean account of motivation or motivating reasons in [63, p. 70].

achieve such forms of the good. The answer proposed by the relevant account of human vulnerability and flourishing is based on a number of observations of contingent features of human nature, several of which will appear in what follows. However, at the very core of these is the observation that human beings from early childhood inevitably question not only why they should act in one way rather than another in a given situation, but also what makes a specific reason for action better than another. To put it slightly differently, it seems as though from early childhood it is an inescapable human character trait to search for and question possible (normative) reasons for action.[35] In terms of answering the question of what it requires to achieve those forms of the good conducive to human flourishing, this observation has clearly paved the way. If human beings are from early childhood searching for and questioning possible (normative) reasons for action, then the achievement of any good conducive to human flourishing is conditional on the ability to find reasons for action that, when they result in action, will achieve forms of the good conducive to human flourishing. More specifically, the achievement of forms of the good conducive to human flourishing is clearly conditional on the exercise of the ability to critically reflect on, evaluate and justify possible (normative) reasons for action: i.e., the ability of practical reasoning.[36] This ability appears to have certain prerequisites.

First, as (normative) reasons for action may be thought of as propositions stating what is desirable or required in a given situation, the reflection and passing of judgements on reasons for action seems to require the mastery of language.[37] Second, as human beings are seemingly directed from birth only at those forms of the good constituted by the satisfaction of immediately felt bodily desires, the very ability to evaluate different possible reasons for action also requires the ability to recognize forms of the good other than those directed from birth.[38] Thus, in stating that something is desirable or required, a (normative) reason for action is clearly defined by some form of the good that an action may achieve. Third, evaluating possible reasons for action clearly also requires the ability to stand back or detach oneself from immediate desires in the sense of being able not to let these desires play a privileged role

---

[35] MacIntyre in [63, pp. 66–68]. The reasons MacIntyre is referring to are clearly what we termed normative reasons in Chapter 3.

[36] Macintyre in [63, pp. 67, 77 and 105].

[37] MacIntyre in [63, pp. 53–54].

[38] MacIntyre in [63, pp. 71–72 and 93ff.].

in the reasoning about the good to be pursued.[39] This is simply a conceptual feature of the concept of evaluating reasons for action. Fourth and finally, at a given point in time there seems to be, at least in the minds of human beings, several possible future courses of the world. Since such different alternative futures may present different achievable forms of the good, the evaluation of possible reasons for action clearly also requires the ability to imagine different possible futures and the ability to attach, in a rough and ready way, fairly accurate probabilities to their possible actualization. This further requires not only knowledge of oneself and of the surrounding social and natural world but also sufficient imagination in order to project how the future may come to encompass different opportunities for the unfolding of the self.[40]

According to the account of human nature we are exploring here, the listed abilities underlying the ability of practical reasoning are not innate. That is, the abilities underlying the ability of practical reasoning are neither things a child has from birth nor things the child has the capacity to develop on its own. Rather, it is a basic claim of this position that human beings at birth are at the level of other animals, where this is taken to mean that human beings at this stage do not possess the abilities underlying practical reasoning.[41] In accordance with this view on the animality of human beings, it is furthermore the contention of this account of human nature that the abilities underlying the ability of practical reasoning can be developed and sustained only in and through social relationships.

To be more specific, the claim here is that the ability to recognize various forms of the good and the ability to detach oneself from immediate desires require that parents from childhood promote such skills in the child. However, it is suggested that this may happen by involving the child in co-operative activities – practices – to which both external and internal goods may be attached, where an external good is a good that may also be achieved through participation in other practices and an internal good is a good that may only be acknowledged and achieved through participation in a particular practice.[42] Playing chess is an example of a practice in which the external goods are money and fame and the internal goods are certain game-specific

---

[39] MacIntyre in [63, pp. 72–73 and 88ff.].

[40] MacIntyre in [63, pp. 74–76 and 94–95].

[41] MacIntyre in [63, pp. 4–5, 8, 40–41, 49, 56, 68–69, 155]. However, MacIntyre argues in [63, pp. 55–59] that animals may make rudimentary judgements regarding what they have reason to do based on perception.

[42] MacIntyre in [64, p. 188].

strategic skills. Through persistent participation in a practice the child may learn about different kinds of goods but may also learn to fill out a role or function within the relevant practices independently of immediate desires.[43]

In terms of the ability to imagine different possible futures and the ability to attach probabilities to the possible actualization of these, the knowledge of oneself and of the surrounding social and natural world presupposed by this ability is also something that must apparently be derived through social relationships. Not only in childhood but also later in life, the ability to imagine different possible futures and their probability in the face of a particular state of the social and natural world most often seems to require knowledge that goes beyond the knowledge a person may build through his or her own experiences.[44] However, dependency on social relationships becomes even stronger when we consider how we acquire the self-knowledge that is evidently also a prerequisite of imagining the future course of the world: e.g., knowledge of our temper, skills or physical abilities. Thus, if one condition of the criterion of correct ascriptions of, for instance, identity to a person lies within a community of people,[45] then it seems as though we cannot rely on self-ascriptions of identity: i.e., we cannot rely on ascriptions of properties to ourselves. Hence for us to have genuine and reliable beliefs regarding who and what we are requires that we share beliefs regarding our identity with others for the possible correction of these beliefs. Adequate knowledge of oneself is consequently a shared achievement.[46] Note, very importantly, that, although the claim that the criterion of correct ascriptions of identity lies within a community of people may be contested, it still seems a fairly reasonable – but significantly weaker – claim that, in our building of knowledge of who and what we are, we are to a significant extent dependent on the knowledge other people have of us. On the basis of these considerations of how the abilities underlying the ability of practical reasoning are dependent for their development on and sustained by social relationships, we will now move on to consider, finally, the implications for the possible vulnerabilities of human beings.

Given that human flourishing is dependent on the exercise of practical reasoning as argued above, we are now in a position to clarify further the nature and character of

---

[43] MacIntyre in [63, pp. 89 and 91–92].

[44] MacIntyre in [63, p. 94].

[45] MacIntyre in [63, pp. 94–95]. MacIntyre is here inspired by Wittgenstein in [104, sect. 202ff. and sect. 241].

[46] MacIntyre in [63, p. 95].

human vulnerabilities, of related threats and of human dependency on others. Thus it seems to follow straightforwardly that humans are vulnerable to anything that may inhibit or frustrate both the development in childhood and the sustaining in adulthood of the ability of practical reasoning, and hence of the abilities underlying this one. Since the ability to recognize forms of the good other than the immediately desirable ones and the ability to detach oneself from immediate desires are abilities to be learned throughout childhood, the threat to such development is lack of parental care – the kind of care that on a basic level manifests itself in the protection of the child against threats to human survival and against threats to the development of language and, more generally, basic reason-evaluating capacities.[47] This includes the kind of care that manifests itself in parents teaching their children both the different forms of the good associated with different practices and the persistence required to obtain such benefits, regardless of immediate desires.[48] Since adults may come to suffer from physical and psychological disorders and illnesses, these threats reverberate throughout adulthood. Thus in incurring disabilities an adult may be in need of others to obtain resources needed for survival and to assist in finding new ways forward, for example in recognizing other forms of the good to be achieved in spite of disabilities.[49]

The ability to imagine alternative possible futures and to attach an accurate probability to the possible realization of these is threatened both in childhood and in adulthood and is therefore of special relevance here. One such threat is the failure of others to share their knowledge regarding the workings or laws of the social and natural world. Without others sharing their knowledge about opportunities for achieving certain benefits by participating in communal life and by drawing on the resources of nature and about the risks and dangers involved in the pursuit of these benefits, the number and character of the achievable benefits believed to belong to different possible futures may be wrong. This sharing of knowledge must clearly also include knowledge of the person whose future is at stake. Without self-knowledge a person may simply fail in judgements concerning which futures are realizable given a certain identity or personality.[50] As demonstrated, self-knowledge is a shared achievement, and therefore a person's reasoning about what constitutes goods in different possible futures for that particular person is severely threatened by others failing to share their

---

[47] MacIntyre in [63, p. 72].

[48] MacIntyre in [63, pp. 88–89].

[49] MacIntyre in [63, p. 73].

[50] MacIntyre in [63, p. 95].

knowledge about the identity, character or personality of that person. The threats to
the sustenance of the ability of imagining different possible futures and of attach-
ing probabilities to them may be further extended. Assuming human beings to be
fallible in the sense of being at any point in life susceptible to committing errors in
their practical reasoning because they, for example, lack information, misinterpret
evidence, rely on unsubstantiated generalizations or are prone to moral failures such
as making prejudiced or insensitive judgements about others, there is a clear threat
that these mistakes will remain uncorrected.[51] That is, others will either fail to make
us aware of our mistakes or to raise our awareness of the sources of these mistakes.
This in turn implies that we as humans, perhaps not surprisingly, are threatened by
misdirection, manipulation, exploitation or victimization.[52]

Let us briefly summarize the main results of this analysis of the concept of vulnerability. It was initially established that human beings are vulnerable to or threatened
by anything that may inflict damage on their capacity for practical reasoning. Subsequently it was shown how this capacity for practical reasoning is in turn conditional
on the cultivation of social relationships securing the development and sustenance of
other abilities: namely, those of recognizing forms of the good other than the gratification of immediate desires, of detaching oneself from immediate desires and of
imagining different possible futures through care and the truthful sharing of knowledge. As was noted at the start of this section, there may be more to the notion of
human vulnerability than is outlined here, but the strength of this account is that
it establishes and describes human vulnerability by considering on a meta-level the
prerequisite of achieving what may be regarded by some normative theory as a good.

6.3.2   Belief, reality and vulnerability

The third hypothesis makes two important claims: first, that the strength of our belief
in the reality of a particular patient with whom or which we are interacting is linked
to a belief in the patient being vulnerable to or dependent on our actions; and second,
that our belief in the vulnerability or dependency of the patient is linked to our having
some kind of evidence of the vulnerability or dependency of the patient.

As already stated in previous passages on the first and second hypothesis, there
seems to be no *prima facie* reason to suppose that the belief in the vulnerability or

---

[51] MacIntyre in [63, pp. 95–97].

[52] MacIntyre in [63, pp. 97–98].

dependency of the patient is formed in a way that is different from the formation of other beliefs about the patient. Hence the second claim may be taken to assert that we generally form beliefs about the patient on the basis of what we consider to be relevant and reliable evidence. In that sense the second claim is shared with the first and second hypothesis and will therefore be dealt with in the section following this.

The remainder of this section is therefore dedicated to an analysis and discussion of the first claim. The analysis and discussion is spilt into two parts. First, the logical character of the first claim of the hypothesis is revealed. Second, an attempt is made to support the hypothesis in and through the consideration of an imaginary case.

Believing in vulnerability

The first claim of the hypothesis needs elaboration on two important points. First, we need to clarify the content of an agent's belief in the patient's vulnerability to or dependency on the agent's actions. To do so, we will make use of the theory of human vulnerability and dependency introduced above. In spite of its specific focus on the prerequisites of practical reasoning, it may be helpful in giving both a minimalist and an uncontroversial content to our everyday beliefs in the vulnerability and dependency of other people. Second, the logical relationship between the belief in the vulnerability and dependency of the patient and the belief in the reality of the patient also needs clarification.

In terms of clarifying an agent's belief in the patient's vulnerability to or dependency on the agent's actions, let us begin by noting that the theory introduced above may help us in giving more specific content to an agent's belief in the patient's vulnerability to or dependency on the agent's actions. As such, the theory may be employed in at least two different ways. On the one hand, it may be taken to present a picture of certain elements of the basic human condition that correspond to the common conception of what it means to be human. That is, the theory may be assumed at least in part to reflect how, in our everyday lives, on a very basic level we conceive of other people as human beings. Used in this way, the theory clearly implies that we basically conceive of our fellow human beings as being inevitably caught up in practical reasoning regarding the good to be pursued, and in the exercise of this reasoning and in the pursuit of the good they are vulnerable to and dependent on the input of others. On the other hand, it may also simply be used as an account of the human condition, which implies a conception of human vulnerability and dependency that is not controversial in the sense of necessarily involving substantial normative claims regarding a commonly shared conception of the good of mankind and that is minimalist in the sense that it does not take into account the substantial normative beliefs regarding the good for mankind that each of the agents in the world may entertain.[53] Although the first way of using the theory seems to rest on fairly reasonable assumptions regarding our basic conceptions of each other as human beings, we will here put the theory to use in the latter way. That is, we will use the theory simply as an uncontroversial and minimalist way of deepening the concept of human vulnerability or dependency as it was initially introduced. This is done below by outlining different aspects of a patient's vulnerability and dependency that may be part of an agent's beliefs when engaging in everyday interaction with others.

The first aspect of the patient's vulnerability has to do with the patient's physical and psychological health or well-being. Thus a patient is vulnerable in the sense that, as a consequence of the actions of others, the patient may be injured, disabled, suffer illness or lack resources vital for the preservation of health or well-being. As such, the patient's vulnerability may be related to our actions both positively and negatively as well as directly and indirectly. We may act in ways that affect the patient's health or well-being both directly and indirectly: for example, by intentionally passing on flu by sneezing in the face of another person or by stealing the money they need to cover living expenses. However, we may also omit an action where this omission affects the patient's health or well-being both directly and indirectly: for example, by intentionally refraining from covering our mouth when sneezing or from assisting a person when they are being robbed of their money. The second aspect of the vulnerability of the patient has to do with the patient's exercise of their cognitive or intellectual powers. Thus patients are vulnerable in the sense that they may come to hold false beliefs, reason falsely, be too influenced by idiosyncratic sentiments in their judgements or simply lack the knowledge required in order to make informed choices as a consequence of their actions. As such, the patient's cognitive or intellectual vulnerability may also be related to our actions both positively and negatively as well as directly and indirectly. We may act in ways that affect the patient's exercise of cognitive or intellectual powers both directly and indirectly: for example, by intentionally passing on the incorrect information that all swans are white to the patient or to someone who is a source of information for the patient. However, we may also omit action where

---

[53] For the sake of completeness it has to be noted, though, that the MacIntyrean conception of human vulnerability and dependency rests on the normative claim that it is of value to develop and maintain the ability of practical reasoning.

this omission affects the patient's exercise of their cognitive and intellectual powers directly and indirectly: for example, by intentionally refraining from correcting the patient's false belief that all swans are white or from correcting the same belief held by someone serving as a source of information for the patient.

Having uncovered two possible aspects of a patient's vulnerability, we may now conclude this investigation of the possible content of an agent's belief in the patient's vulnerability and dependency. The starting-point was the formulation of the hypothesis that an agent is convinced of the reality of a particular patient to the extent that the agent comes to believe the patient to be vulnerable and dependent. In the light of the two aspects of human vulnerability and dependency revealed above, we may now rephrase the hypothesis the following way: agents are convinced of the reality of a particular patient to the extent that they come to believe that they may act or omit action in ways that will affect the patient's physical or psychological health or well-being or the patient's sound exercise of their cognitive and intellectual powers. This rephrasing of the hypothesis seems to have brought clarification of an agent's belief in the patient's vulnerability. We will now turn to the second point that needs clarification: the logical relationship between the belief in a patient's vulnerability and dependency and the belief in the reality of the particular patient.

So far in our analysis of the concept of the patient's vulnerability and dependency we have focused entirely on the special case of the patient being another human being. In our analysis of the first and second hypothesis we investigated the relation between the belief in the determinateness of the patient and the reality of the particular patient as well as the relation between the belief in causally affecting the patient and the belief in the reality of the particular patient. In both cases our investigation was conducted on the assumption that a human being was only one possible instance of a patient among others. Although vulnerability in the overall sense of being potentially harmed or damaged may apply to a wide range of entities, we will continue our inquiry into the relationship between the belief in the patient's vulnerability and the belief in the reality of the particular patient in line with the preceding investigations by focusing on the special case of the patient being another human being. Apart from being consistent with our considerations throughout this section, this choice is also founded on the apparent difference in the role of the vulnerability of an entity for our belief in the reality of that particular entity. Thus it seems reasonable to suppose that the vulnerability of another human being is very important for our belief in the reality of that particular human being, whereas it may seem less plausible that the vulnerability of a $100 bill is important for our belief in the reality of a $100 bill.

Given that the hypothesis claims something regarding the particular class of patients being co-extensive with the class of human beings, the intention is that it should make a rather strong claim: namely, that the belief in the vulnerability of the patient is a necessary prerequisite of the belief in the reality of the particular patient. The underlying assumption is that the vulnerability of another human being is generally believed to be such a fundamental feature of human existence that it is impossible to be, and hence exist as, a human being without actually being vulnerable and dependent. However, believing in the vulnerability and dependency of someone is clearly not sufficient for the formation of a belief in the existence of another human being. Thus we may come to believe that Achilles was indeed vulnerable without necessarily believing Achilles ever to have existed. With this clarification of the logical relationship between the belief in the vulnerability and reality of the patient we will move on to consider an argument in favour of this relationship.

The mannequin argument

The main argument to be advanced here involves life-size tailor's dummies, or mannequins. Before launching this argument let us briefly consider the *prima facie* credibility of the first claim of the third hypothesis.

On the face of it, the claim seems quite reasonable. It seems as though in our everyday dealings with other people we are more or less constantly faced with, aware of and acting in response to the vulnerability and dependency of other people. For example, driving a car seems to involve a more or less constant awareness of the vulnerability of other groups of road users. Likewise teaching or the supervision of fellow employees in most cases also seem to involve an awareness of their vulnerability to the words and knowledge of the teacher or supervisor and so on. Unfortunately, these examples are not at all conclusive in terms of establishing the relevant link between our belief in the vulnerability of a person and the reality of that person as a particular person. Granted that the examples show that we are aware of, think about or believe in the vulnerability of those people with whom we are interacting in our daily lives, this does not mean that we generally take vulnerability to be a necessary prerequisite of human being. Thus it is fully consistent with these examples that we come to ascribe the property of vulnerability to our fellow human beings as some sort of accidental or non-essential feature. Ultimately then, although the examples may be indicative of our association of vulnerability and human being, they do not vindicate the first claim that our belief in the particular human being is inextricably linked to human vulnerability. In order to show this, we will here use an argument involving the concept of a mannequin.

Imagine that our Olympic javelin-thrower has survived his ill-fated journey through the Sahara and after a lengthy period of recuperation is to be found walking down the high street of his home town on a snowy winter's day. As he passes a large shopping centre, his attention is caught by a row of extremely lifelike mannequins placed in the window on the first floor of the shopping centre. At least one of the mannequins has very distinct facial lines and a vivid facial expression, and the javelin-thrower asks himself whether the mannequin is a real human being. Suppose that, his curiosity being too strong and his skills of throwing too useful, he cannot resist the temptation to seek a definite answer to his question and therefore makes a snowball, which he forcefully directs at the part of the window right in front of the figure in question. As the snowball hits the window with utmost precision, the figure behind the window reacts with a rapid movement of the head.

Yet again, a few comments may be helpful for the understanding of the interpretation of the example that will be suggested below. First, the setting is chosen because it seems as though the question raised in this situation is exactly the question of whether or not a particular entity bearing a striking resemblance to a human being is really a particular, living human being. Second, it seems as though the javelin-thrower clearly believes that the existence of a particular entity is affected by his throwing of a snowball. Taking into account the second hypothesis, his belief in the reality of the particular entity at which he directed the snowball may be founded on his belief in having causally affected the particular entity, but perhaps also on his belief that the effect on the particular entity is constitutive of his life-world.

The case of the javelin-thrower and the mannequin may come to count in favour of the first claim of the hypothesis on the basis of three suppositions all enjoying a certain *prima facie* plausibility. First, it seems plausible to suppose that the javelin-thrower, faced with the relevant evidence of having been the cause of the movement of the figure bearing a striking resemblance to a human being and lined up alongside the mannequins, will believe this entity to be a real and particular human being. Second, it seems plausible to suppose that the reason the javelin-thrower comes to believe the entity to be a human being, and not just a mannequin, is in part simply the nature of the effect on the particular entity in question. Moreover, it seems as though his belief that the figure is a real and particular human being is, at least in part, based on his belief that the actual effect on the entity corresponds to the nature of the possible effects his action would have on a human being. Third and finally, it seems plausible to suppose that the javelin-thrower's belief that the entity reacted to his act of throwing a snowball at it in a way that human beings would normally react to such actions reflects an underlying general belief in the vulnerability and dependency of human beings. More specifically, it seems as though the correspondence believed to hold between the act of throwing a snowball at a human being and the reaction of that human being to this threat reflects an underlying belief in humans being constituted in a way that make them vulnerable to the effect of a number of objects impacting them; i.e., it reflects a more general belief in humans being vulnerable in terms of their physical health and well-being.

Given the plausibility of the four suppositions made above, the example of the javelin-thrower and the mannequin clearly serves its purpose. The aim of this section was to ground the claim that an agent's belief in the patient's vulnerability and dependency is a necessary prerequisite of believing in the reality of the particular patient. The case of the javelin-thrower and the mannequin simply shows that the belief in the patient's vulnerability and dependency is used as a criterion of an entity's being human. Thus it is, at least in part, because of the recognition of vulnerability exhibited in the patient's rapid move of the head that the javelin-thrower comes to believe in the reality of the entity as a particular human being. Obviously other features of the situation and of the figure strikingly similar to a human being may have contributed to the belief in the reality of the entity as a human being, such as features contributing to the javelin-thrower's belief in the determinateness of the patient or the causal effect on the figure, and consequently the belief in the vulnerability of the patient may not be sufficient for the belief in the reality of the figure as a particular human being. However, the case still seems to support the idea that the belief in the patient's vulnerability and dependency is a prerequisite of the belief in the reality of the patient as a particular human being.

The interpretation of the case of the javelin-thrower may be countered on two important points. The first is the generality of the claim that vulnerability is always a necessary prerequisite of a belief in the reality of a particular patient, and the second the generality of the claim that it is vulnerability in general that is a prerequisite of the belief in the reality of a particular patient. In terms of the former, the objection may be refuted by the fact of vulnerability seemingly being used as a criterion of human existence in the case under consideration. Thus there seems to be very little to suggest that the relevant association of the vulnerability and the reality of the particular figure in question is restricted to the particular situation. On the contrary, it seems a reasonable interpretation that the javelin-thrower in this situation draws on a belief in the relation between the vulnerability and the reality of a human being that is general in nature. The latter objection may at first sight seem more to the point. Strictly speaking, the case of the javelin-thrower and the mannequin does not support the notion that the javelin-thrower takes physical vulnerability in general to be a prerequisite of the reality of the particular entity as a human being – he may believe vulnerability to snowballs in the facial region to be a prerequisite of only a particular human being. The answer to this objection is simply that the central point of the case could have been illustrated in a number of ways. Thus the javelin-thrower could have thrown a stone directed at the torso of the relevant entity, or he could have used a flame-thrower. It seems as though all such illustrations would point to the conclusion that the underlying belief in the vulnerability of the entity in question is a general belief in the physical vulnerability of the entity. In fact, the case may actually be extended to show that the javelin-thrower's belief in the patient's vulnerability and dependency also has a psychological aspect.

Imagine that the figure lined up next to all the mannequins bears a striking resemblance to the hated ex-wife of the javelin-thrower. Torn between sheer curiosity as to whether or not it really is a living human being – and possibly his ex-wife – and the desire to express his strong sentiments regarding her personality, the javelin-thrower aggressively runs into the adjacent bookshop and returns with a huge poster, on which he carefully writes a strongly worded and abusive message to his ex-wife. He then attaches the poster to a long pole and resumes his position outside the shopping centre waving the poster in front of the window behind which the figure who looks strikingly similar to his ex-wife is standing. Leaving this extended version of the case of the javelin-thrower and the mannequin at this point, the central question to be asked and answered is whether it seems likely or reasonable that one would believe it possible in principle to settle the question of the reality of the entity in the window using the method of the javelin-thrower. In short, we have to address the question as to whether or not it seems reasonable to suppose that the figure might react to such a poster. If this is granted, then it seems as though the underlying belief regarding the vulnerability of the entity also encompasses vulnerability to the abusive words of others: i.e., some sort of psychological vulnerability.

6.4   Hypotheses I to III: Beliefs and evidence

As promised in the previous sections of this chapter, the claim shared by all three hypotheses concerning the way in which human beings form beliefs also requires scrutiny.

6.4.1   Linking beliefs and evidence

As the concept of evidence was elaborated on previously, we can enter straight away into the discussion of the claim shared by all three hypotheses. This is done in this section, consisting of three subsections. In the first of these the logical character of the common claim of the hypotheses is uncovered. In the second the attempt is made to support the hypothesis in and through the consideration of the role of knowledge for human intentional behaviour. In the third the claim is considered in the light of the special context of interaction, and it is argued that at least in this context the claim shared by all of the hypotheses seems to hold.

Evidence preconditioning belief

In the explication of the claims of the hypotheses in preceding sections we noted how they seem to share the claim that we generally form beliefs about the patient on the basis of what we consider to be relevant and reliable evidence. Let us briefly examine this claim.

The four cases studied in some detail in the previous three sections – i.e., the case of the joker, the javelin-thrower and the oasis, the stockbroker and the javelin-thrower and the mannequin – all seem to provide some support for the claim that the formation of the beliefs underlying the belief in the reality of a particular patient involves evidence. Thus they are all examples of exactly how the beliefs underlying the belief in the reality of a particular patient are formed on the basis of evidence in a concrete situation of interaction. Although true, this observation rests, however, on a mistaken interpretation of the claim shared by all the hypotheses. The claim is not supposed to be interpreted as holding that the formation of beliefs about the patient of our actions always involves evidence as a coincidental matter of fact, but rather that our formation of beliefs presupposes evidence. Thus the claim shared by the hypotheses is to be interpreted in the stronger sense according to which evidence of the belief underlying the belief in the reality of a particular patient is generally a prerequisite of coming to hold the relevant belief. That is, generally we require evidence for something being the case, say evidence that p, in order to form and hold the belief that p – and the more evidence we have of p being the case the more certain or convinced we are of our belief that p.

The stronger interpretation of the claim shared by all of the hypotheses has to do with the arguments to be launched in the next chapter. Here we will be arguing that the conditions of interaction in cyberspace are such that they reduce the availability of reliable and relevant evidence. Clearly, without the strong interpretation of the shared claim, this loss of evidence may not imply the loss of any of the beliefs underlying the belief in the reality of a particular patient, and thus by extension of the belief in the reality of a particular patient.

Arguments based on dependency upon knowledge

At the core of the shared claim of the hypotheses lies the more general claim that in our formation of beliefs we require evidence of the believed. That is, we believe that p only if there is evidence of p being the case: e.g., we believe the coffee to be strong only if there is evidence of it being strong, or we believe there to be birds in our garden only if there is evidence of there being birds in the garden and so on. A possible argument in favour of this claim runs along the following lines. We, as human beings, generally strive to transform our beliefs into knowledge. This process is normally taken to require the justification of our beliefs, where the justification of our beliefs is a matter of supporting them by means of evidence. Hence we are generally striving to support our beliefs by means of evidence. Let us briefly deal with the different aspects of this argument.

To take, first, our urge towards knowledge, one possible source of this may be our need to make informed choices in order to achieve our goals or purposes in and through action. It seems to be a constituent of normal human beings that we entertain goals or purposes to be achieved in and through action. A large proportion of the goals or purposes we entertain are of such a nature that they require certain specific actions in order to be achieved (most fully), and hence they require that we decide to act in one way rather than another. To decide the way to act in order to achieve a certain goal clearly involves and requires the formation of beliefs regarding the state of the world and the possible means-end relations within it: i.e., the formation of beliefs regarding which actions would be efficient in achieving the goal. At this point the possession of true beliefs becomes of vital importance since we may form various beliefs about the world and our means-end relations within it – what matters, however, is that these beliefs are true in order that our choice of action will lead to the achievement of the relevant goal. Knowledge is usually taken to be constituted in part by true beliefs, and thus it is ultimately knowledge that comes to be of vital importance. Hence our quest for knowledge.

Second, the argument rests on our association of knowledge with justification. Given that, at least in situations in which we are to act on the basis of a goal or purpose, we are looking for true beliefs, and ultimately knowledge, regarding the world and its workings, how do we attempt to distinguish those of our beliefs that are true from those that are untrue? It seems as though in our everyday dealings we normally try to make this distinction on the basis of justification. That is, we try to justify our beliefs in the sense of giving reasons in support of the content of the belief being the case – and the more and the better reasons we have in support of a given belief, the more likely we are to believe it to be true. Hence, phenomenologically speaking, justification seems to be our way of bridging the gap between beliefs and knowledge.

Third, the argument identifies justification as supporting a given belief by means of evidence. We have just defined justification as giving reasons in support of the content of a belief being the case. It should be evident that this involves supporting one belief by means of other beliefs. In the previous chapter we defined evidence as supporting a hypothesis by means of beliefs. Hence when we justify a belief, we are actually supporting it by means of evidence.

The argument introduced at the outset of this section may be recapitulated in the light of our investigation in terms of the well-known definition of knowledge as justified true beliefs (JTB). Thus the argument claims that for various reasons we generally strive to form beliefs that qualify as knowledge defined as (JTB). In other words, we generally only form the belief that p on the basis of having evidence in favour of p being the case.[54]

Arguments based on special contexts

Although in the preceding paragraphs we argued that the second claim of the hypothesis really reflects a more general approach to our formation of beliefs, it is worth noting that the claim shared by the hypotheses actually concerns patients with whom or which we are interacting. Thus none of the hypotheses as initially stated actually

---

[54] Gettier in [33, pp. 121–123] shows that JTB does not warrant knowledge. This need not bother us, however, as the claims advanced here concern our everyday practice of building knowledge.

rules out the possibility of coming to believe p without having evidence of p being the case. Hence they leave room for the possibility that the requirement of evidence only applies to or is significantly stronger in the context of interaction. We may perhaps believe in the determinateness or the vulnerability of other entities, for example, without having any evidence of this but simply because we are not required to interact with them. They simply do not concern us or concern us less than patients with whom or which we are to interact. Thus we may perhaps believe in the determinateness and vulnerability of life on other planets without having any direct or indirect perceptual evidence of this being the case simply because we are not required to interact with them.

In order for there to be such a difference between our formation of beliefs regarding a patient with whom or which we are interacting and the formation of beliefs regarding an entity with whom or which we are not interacting, it seems as though there would have to be a relevant distinction between the two cases apart from the fact of the one comprising interaction. One such distinction could be that in interaction we are by definition engaged in intentional action: i.e., action carried out for a reason or with a goal or purpose. As argued in the previous paragraph, intentional action may seem to place special requirements on our formation of beliefs, whereas the formation of beliefs regarding entities with whom or which we are not interacting do not seem to impose such requirements on us, since we are not dependent on the truth of the beliefs in order to achieve a certain goal.

We will not pursue this line of reasoning further but simply note that these considerations of the context of interaction seem to introduce a way of salvaging the claim shared by all three of the hypotheses in the event that the argument based on the human dependency on knowledge is contested.

Chapter 7

Belief and evidence

This chapter continues our attempt to develop a model that will explain the ethical difference between interaction inside and outside cyberspace captured in the basic premise ([TBP]). In the previous two chapters we introduced and investigated three hypotheses linking an agent's conviction concerning the reality of a particular agent to certain other beliefs and to the availability of evidence in support of these other beliefs. That is, we covered the first stage (see the first section of Chapter 5) of our attempt to explain the basic premise: we showed how agent A's belief that p is dependent upon the conditions C.

In this chapter the second stage is covered. Here it will be shown how the conditions of interaction in cyberspace limit the availability of evidence in support of the beliefs underlying the belief in the reality of a particular agent. In the terminology used above, it will be shown that the conditions C are not satisfied in particular kinds of interaction in cyberspace.

The chapter is crucial for our attempt to provide an alternative account to that given by Levinas of the role played in ethics by the face of another person. Thus by showing how the evidence needed for the formation of certain beliefs is lost in cyberspace, we will have taken a crucial step in clarifying the role for ethics of the face of another person.

T. Ploug, *Ethics in Cyberspace: How Cyberspace May Influence Interpersonal Interaction,* **157**
© Springer Science+Business Media B.V. 2009

7.1   Evidence in cyberspatial interaction

Having elaborated on the three hypotheses in the previous chapters, it is now time to apply them to the case of interaction in cyberspace as exemplified in text-based chatrooms and tele-operation. We will do this by considering to what extent it is possible to gather evidence about the patient in these kinds of interaction in cyberspace.

These considerations will draw on the three key properties of interaction in cyberspace outlined in Chapter 4: the limited exchange of data and information in cyberspace, the limited perceptual access between the agent and patient, and the anonymity of agent and patient in cyberspace. And, as already anticipated, our considerations will show that in all interaction in cyberspace there is less evidence of the interacting party available than in face-to-face interaction, and that in the particular kinds of interaction the available evidence is less relevant and reliable than the evidence obtained in face-to-face interaction.

7.1.1   Lack of evidence

At the most fundamental level the decrease in the accessible or available evidence about the interacting party or patient has to do with the limited exchange of data between interconnected computers. As has already been argued in the subsection on the key properties of interaction in cyberspace, this limitation is also a limitation on the information exchanged. The obvious challenge now is to show that the limitation on the exchange of information is also a limitation on the exchange of evidence.

The relation between information and evidence seems to be fairly straightforward. In the subsection on the key properties of interaction in cyberspace, information was defined very briefly as meaningful data, and immediately above we defined evidence as a set of observational beliefs playing the role of supporting a hypothesis. Given these definitions, it seems to follow that to have evidence in support of it being the case that p, in at least certain cases, involves having information to the effect that p. Suppose, for example, that B's evidence in support of the hypothesis that person A is angry consists in B's belief that person A recently uttered a statement to the effect that she is angry, and that B formed this belief on the basis of hearing person A shout 'I am angry'. Conclusive or not, B's evidence in support of it being the case that person A is angry clearly involves B being in possession of information – i.e., meaningful data – to the effect that person A is angry. In the example the data may be said to be the syntactically well-formed words shouted by A, and the meaning

is then determined by the semiotic system attached to natural language. The point
we are trying to make here regarding the building of evidence must be distinguished
from a related point regarding the formation of knowledge. Thus the point made here
is that the building of evidence requires the availability of information, whereas one
may reasonably claim that the building of knowledge requires true information also
known as factual information.[1]

Since we are here interested in the conditions surrounding interaction in cyberspace, we may side-step the more general issue of whether or not the building
of evidence always requires information. In the case of interaction in cyberspace or
computer-mediated interaction, therefore, the building of evidence always requires
information. Given the workings of a computer, interaction in cyberspace is simply
conditioned by the exchange of well-formed data that in the end are represented in
ways that are ultimately meaningful to human agents – such as words or pictures
on a monitor. As a consequence, the limitations on the exchange of information are
obviously also limitations on the availability of evidence in interaction in cyberspace.

In order to understand fully the significance of this result, let us make a few clarifying remarks. First, it is worth noting that the limitation is a limitation independent
of general human shortcomings, inabilities or fallibilities, and independent of other
relevant limitations of the material world. In short, it is a limitation constituted by the
difference – regardless of how it may be specified – between computer-mediated interaction and face-to-face interaction. That is, the limitation can be dismissed neither on
the grounds that we as humans are capable of receiving and processing only limited
information because of, for example, limited cognitive or intellectual powers. Nor can
it be dismissed on the grounds that in face-to-face interaction the limitations of matter
also apply. Thus the point made here is that, when considered as a space or place for
interaction, the conditions of cyberspace are such that there is less available information than in face-to-face interaction. Before our presence in cyberspace – i.e., before
the human condition enters the scene – there is simply less available information than
in the situation of being face-to-face. Furthermore the limitations of matter – e.g.,
the limitations of the human sensory organs – that apply to face-to-face interaction
clearly also still apply to interaction in cyberspace since in this kind of interaction
we are not simply connected directly to the brain or mind of another agent: i.e., we
are not in any way relieved of the limitations of the materiality and physicality of the
human body.

---

[1] Floridi in [25, p. 46]. Cf. also Adams in [1, p. 229].

The main upshot of these considerations is that in cyberspace there is not the richness of information – and therefore of evidence – that will be found in face-to-face interaction. Through a variety of technological equipment information about the interacting parties can be made available. However, there is always a limit to the information and evidence available, and this limit is constituted by the information and evidence being exchanged by means of interconnected computers. To illustrate the point in a more concrete manner, let us relate the limitations of cyberspace to the technological equipment of those particular kinds of interaction in cyberspace with which we are concerned here. Suppose that after a while one of two people having a trivial conversation in a text-based chat-room is asked to provide a description of the cup she is holding in her hand. By way of description she says, 'the cup has coffee in it'. This description obviously provides very little information regarding the cup, the coffee or the coffee in the cup. This flaw may in part be remedied by getting visual access to the person with whom the interaction is taking place. This will, however, be only in part, since there will still be information regarding the cup and coffee that will not be available, such as the smell and taste of the coffee.[2] Without going into further detail, the point of this illustration remains the same as above: namely that, regardless of how much technological equipment is brought into play, the richness of information and evidence will always be limited in cyberspace simply because of the limitations of the technological equipment: i.e., the interconnected computers by which it is exchanged.

At this point it seems reasonable to ask whether the loss of information and evidence in cyberspace is at all relevant – and what we should we make of any loss of information and evidence. The imagined possibility underlying these questions is that the loss caused by the exchange of information and evidence occurring in cyberspace is of information and evidence irrelevant for human purposes. In that case the loss would not have any real significance for human agency *per se* – it would simply be of practical irrelevance. Even more pressing than the general consideration of the relevance of the information and evidence lost in cyberspace is the particular consideration at the heart of this subsection, namely whether the information and evidence lost in cyberspace are relevant for establishing the determinateness and vulnerability of the patient and for establishing the causal relations between the agent and the patient. If not, then the loss of information and evidence in cyberspace clearly does not influence an agent's belief in the reality of the particular patient.

---

[2] Dretske in [20, p. 137] for the example of the coffee cup. Cf. also Borgmann in [8, p. 95].

7.1.2   Lack of relevant evidence

The second key property of interaction in cyberspace was the limited sensory access exemplified in both kinds of interaction with which we are concerned here. Let us briefly recapitulate how the relevant kinds of interaction limit sensory access and then try to determine whether any implied loss of evidence is a loss of evidence relevant for establishing the determinateness and vulnerability of the patient or for establishing the causal relations between the agent and the patient. In doing this, we will also have taken a first step in answering the more general question of whether cyberspace may at all limit the availability of information and evidence relevant for human purposes.

In the section on the key properties of cyberspace we saw how interaction in anonymous text-based chat-rooms implies an extensive loss of sensory access to both the patient and the patient's environment. This was shown to involve the loss of access to the properties of the patient's body, such as size and shape, scars and handicaps, wrinkles, clothing, hair, skin colour, smell, gesture, facial expression, decoration (e.g. jewellery), timbre of voice and so on. Furthermore it also involves the loss of sensory access to properties such as colour, temperature, number, shape, size, lightness or darkness of those elements constituting the setting or environment: for example, architecture, natural surroundings, furniture, climate, icons and so on. Although the case of tele-operation is a little more complex, it was also shown in the relevant section that the examples involve the agent having less sensory access to the tele-operated entity, the patient and the patient's environment. A limited number of cameras and microphones may also limit appropriate visual and audio access, and furthermore these kinds of interaction do not incorporate olfactory and tactile access to the tele-operated entity, the patient and patient's environment. Hence there will be a loss of access to the properties of these along the lines of the loss involved in text-based chatting illustrated in the subsection on the key properties of cyberspace and cyberspatial interaction.

As should be clear, both kinds of interaction in cyberspace imply the loss of evidence about the patient since they both restrict the sensory access to the patient, and evidence is defined as beliefs formed on the basis of perception, where perception involves the use of senses. However, the central question to be answered here is whether the loss of evidence is of relevance for the agents' belief in the determinateness and vulnerability of the patient, for the agents' beliefs in having a causal effect on the patient and for the agents' belief in the role of this effect on their life-world. Before answering this question, let us briefly make a rather trivial observation about the general relation between evidence and beliefs, which may, however, give some insight into the strategy pursued below in answering the question.

It seems to be the case that in a number of situations we form beliefs by making deductive inferences from beliefs arrived at by combining evidence with certain background beliefs, where evidence is taken in the earlier defined sense of being beliefs derived from observation. Thus we may more or less explicitly form the belief that the person sitting next to us is a woman on the basis of having evidence of the person wearing make-up in combination with the belief that we are in an everyday situation and the belief that, if this is an everyday situation, then if someone is wearing make-up, they are a woman – where the latter two beliefs may themselves be the outcome of inferences from yet other beliefs. Note that, according to some definitions of justification, a certain belief is justified in the event that it may be inferred from a certain set of beliefs,[3] although we are at present only making the claim that inferences are made, not that these inferences are justified. Given that we form beliefs in this way, it seems that an appropriate strategy in trying to answer the question of whether or not the loss of evidence in cyberspatial interaction is of relevance for the agent's formation of the beliefs in question would be to sort out whether or not the loss of evidence for certain less defining properties of the particular patient may lead to the loss of belief in more defining properties of the particular patient and in turn to the loss of the belief in the determinateness and vulnerability of the patient, the belief in having a causal effect on the patient and the belief in the role of this effect on the agent's life-world. In the following three sections – one for each of the hypotheses – this strategy will be pursued.

Determinateness and relevant evidence

In order to decide whether or not the loss of evidence about the patient in cyberspatial interaction is a loss of evidence relevant for establishing the determinateness of the patient, we will conduct a twofold investigation. First, we will clarify how an agent may arrive at a belief in the determinateness of the patient and thus clarify what evidence is relevant in so doing. Second, we will further elaborate on the consequences of the loss of sensory access to the patient by providing examples of the inferences usually made on the basis of the evidence of the patient.

---

[3] This formulation should be compatible with both a foundationalist and a coherentist conception of justification. Cf. Dancy in [17, pp. 53ff. and 110ff.].

The belief in the determinateness of the patient was defined as holding for any property that at any time in the past or present was either possessed or not possessed by the patient. Put slightly differently, it is the belief in the patient being in this sense a complete person. In assessing how the agent may arrive at this belief about the patient, it seems reasonable to suppose that evidence of those properties of the patient which are indicative of the defining properties of the patient may play a certain role. Since the belief in the determinateness of the patient clearly entails the belief that it holds for the particular property p – for example, being a lover of food – that the patient either has it or not, and since it seems reasonable to suppose that the agent may *in part* arrive at the belief that the patient has the property p by means of evidence of other properties of the patient or of the patient's environment believed to be indicative of whether or not the patient has the property p, then it is clearly the case that the agent may *in part* arrive at the belief in the determinateness of the patient on the basis of evidence indicative of the properties of the patient and setting – provided that the number of properties relevant for the formation of a belief in the determinateness of the patient are finite. It may be the case that the belief in the determinateness of the patient is formed by induction, in such a way that evidence indicative of a number of properties relevant for establishing the determinateness of the patient may lead the agent to believe that it holds for any further property that the patient either has it or not. It may also be the case that the belief in the determinateness of the patient is formed by deduction. If, for instance, the agent holds the belief that certain cardinal properties such as age, gender or race are indicative of the determinateness of the patient, then evidence indicative of these cardinal properties may lead the agent to believe that the patient is determinate. No matter how the belief in the determinateness of the patient is formed, it seems to hold – and this is the crucial point to be made here – that the agent may arrive at this belief on the basis of evidence of the properties of the patient and the patient's environment. With this in mind let us finally consider the conditions prevailing in the gathering of evidence in cyberspatial interaction.

The loss of sensory access to the patient's body as well as the setting implies a loss of evidence about many features that would under normal circumstances be taken to be indicative of the defining properties of the patient, such as the patient's age, gender, race, values, personal history, aesthetic taste, financial situation, psychology, marital status, emotional state, religious and political beliefs and affiliations, family status and so on. Thus the loss of sensory access implies, as already mentioned, the loss of evidence of, for instance, the patient's facial decoration and wrinkles, which would under normal circumstances be taken as rough indications of the gender and age of the patient, but which would occasionally also to some extent be indicative of the patient's aesthetic taste and sometimes also of the character of the life lived by the patient. Likewise the loss of sensory access to the patient implies the loss of evidence of, for instance, the colour of the skin of the patient, which would under normal circumstances be indicative of the race of the patient. Finally, we might also point out that the loss of sensory access to the patient also implies the loss of evidence of the type of clothes and ornaments worn by the patient, which may under normal circumstances be indicative of both the aesthetical taste and the financial situation of the patient, their marital status and sometimes even their religious beliefs. The latter may hold for a patient wearing a particular kind of cross around their neck or a scarf covering the face.

These examples of how the loss of sensory access to the patient may imply the loss of evidence of properties of the patient that may in turn be indicative of defining properties of the particular patient such as personality traits could clearly be further elaborated on. For the present purpose of deciding whether or not the loss of sensory access to the patient may imply the loss of evidence relevant for establishing the determinateness of the patient the previous examples will, however, suffice. Thus, to conclude, given that we arrive at a justified belief in the determinateness of the patient this way, it is clearly the case that the loss of evidence indicative of the defining properties of the patient, other things being equal, implies the loss of evidence relevant for the formation in the belief in the determinateness of the patient.

Causal efficacy and relevant evidence

The next obvious step is to investigate whether or not the evidence lost in cyberspatial interaction is also of relevance for the agent's ability to establish having a causal effect on the patient and the influence of this effect on the agent's own life-world. In doing this, we will follow the same procedure as in the previous section. That is, we will start by considering what evidence is relevant for an agent's attempt to arrive at beliefs with the content in question, and then go on to provide examples of how the loss of sensory access to the patient in cyberspatial interaction also limits the availability of evidence relevant for the formation of these beliefs.

The belief held by agents that their actions causally affect the patient was defined as possibly having four components. Let us deal with each of these in turn. First, the sufficiency or invariability component. Part of the content of a belief in our actions being the cause of an effect is that the action is invariably followed by a given effect – and not that the action increased the probability of a certain effect. Since the belief in the succession of one event from another clearly constitutes a central element of the belief in the invariability between the cause and effect, it seems reasonable to suppose that a patient may *in part* arrive at the belief in the effect invariably following the cause on the basis of evidence of the succession of an event involving the patient on an event involving the agent: i.e., on the basis of evidence of a change in the state of the patient or of evidence of a change in the patient's circumstances following the agent's action. This could, for instance, be evidence of a change in the patient's facial expression, such as the appearance of tears, following an utterance of the agent.

Second, the cause-as-condition component. Part of the content of the belief in one's actions being the cause of an effect is that the cause is not the action in itself but rather the combination of the action and relevant conditions. Since the belief in a set of conditions constituting the cause concerns the particular conditions in which the agent is acting, it seems reasonable to suppose that this belief may also *in part* be formed on the basis of evidence of the particular state of the patient and the particular nature and properties of the patient's environment. In the example of the agent verbally affecting the patient the evidence could, for instance, be features of the patient indicative of the patient generally being sensitive and insecure – but could also be properties of the setting or living conditions of the patient indicative of the patient having just lost her job or the like. The combination of properties of the patient and the setting may be taken to form a set of conditions sufficient for the effect to occur.

Third, there is the counterfactual component. Part of the content of the belief in one's actions being the cause of an effect is that the effect would not have occurred had the action not been performed. Once again, since the belief in this counterfactual relationship concerns what would have happened in these particular circumstances had the action not been performed, it seems reasonable to suppose that this belief may also *in part* be formed on the basis of evidence of the particular state of the patient and the particular nature and properties of the environment. In the example of the agent verbally affecting the patient the evidence could, for instance, be the patient being solely focused on the words of the agent, of the patient being cheerful before the agent's utterance and, more generally, of the absence of other factors that could possibly have caused the sadness of the patient at that very moment. The combination of such properties of the patient and the setting may be taken to indicate that the event constituting the effect would not have occurred had the agent not acted in the relevant way.

Fourth and finally, the manipulability component. It seems to be a very ingrained belief that the causal relation is fundamentally of such a nature that the effect may be influenced by manipulating the cause – especially in cases where the cause involves human actions. The belief in the relation of manipulability between cause and effect is related to the third component. Thus if as an agent we have good grounds for holding the belief that an event taken to constitute a certain effect would not have occurred had we not acted in a certain way, then we would also seem to have good grounds for the belief that we may manipulate the effect by manipulating the cause in the sense of bringing about the effect by bringing about the cause. Hence, as evidence in support of the former involves evidence of the particular state of the patient and the particular nature and properties of the setting, then evidence of the latter will also involve evidence of the particular patient and setting. Having thus shown how an agent's belief in having causally affected a patient may trade on evidence of the particular patient and setting, we may now proceed to consider how agents may arrive at a belief in the effect of their actions on the patient being constitutive of their life-world.

The agents' belief in the effect on the patient being constitutive of their life-world was defined as the equivalent of the belief in the effects of their actions on the patient being so important to their ability to fulfil one or more of their purposes in life that they are influential in shaping their life and actions. Since the belief in causally affecting the patient clearly constitutes a central element of the belief in the effect on the patient being constitutive of the agent's life-world, and since we have just shown that the belief in causally affecting the patient may in part be arrived at on the basis of evidence of the particular state of the patient and the particular nature and properties of the setting, we may conclude that the formation of the belief in the effect on the life-world may *in part* be arrived at on the basis of the evidence in question. Suppose that, in the above example of the agent hurting the patient by verbal means, one of the agent's core purposes in life is to avoid causing other people hurt. It seems evident that in the event that the agent comes to believe that hurt has been caused to the patient this would be considered important for the ability to fulfil an important purpose to a degree that the event will come to shape the agent's life and actions, by leading, for instance, to apologies, to memories of the event, to considering the choice of different words in a similar situation in the future and so on. (How the event may shape the agent's life obviously depends on other beliefs and purposes of the agent.) The point to be made here is simply that, in the agent's formation of the belief in the hurting of the patient being of such importance, the evidence relevant is, as has already been argued, clearly the evidence of the effect on the patient: i.e., the evidence of the particular state of the agent after the utterance. With this in mind, let us finally consider the conditions prevailing for the gathering of evidence in cyberspatial interaction.

As should be clear from the previous paragraph, the loss of sensory access to the patient in cyberspace implies a loss of evidence about the state of the agent as well as about the properties of the environment. Thus the loss of sensory access to the patient implies, among other things, loss of evidence about the patient's facial expressions, exclamations, tone of voice, gestures, other bodily movements, colour of skin, bodily smell, clothing and so on, which may all be indicative of the emotional, psychological and intellectual state of the agent. Paleness of the skin may, for example, be presumed to indicate that the patient is in a state of shock, tears to indicate sadness, trembling fingers to indicate nervousness and so on. Likewise the loss of sensory access to the environment of the patient implies the loss of evidence of the basic properties of the entities constituting the setting such as size, shape, colour or smell, which may in turn be indicative of the character of the entities constituting the setting, such as low-quality housing, an ordered room and the 'forces' at work within the setting, such as wind, sun, gravity, physical attraction, social depravation, harassment – 'forces' obviously relevant for the formation of the belief in the cause being a set of conditions and of the belief in the counterfactual relationship between the events constituting cause and effect.

Given that we arrive in part at a belief in having a causal effect on a patient and in this causal effect being constitutive of our life-world as described above, it is clearly the case that the loss of evidence indicative of the state of the agent and of the character of the setting, other things being equal, implies the loss of evidence relevant for the formation of these beliefs.

Vulnerability and relevant evidence

The final step in these considerations of the relevance of the evidence lost in cyberspatial interaction is obviously to investigate whether or not the evidence lost is also of relevance for an agent's ability to form the belief in the vulnerability and dependency of the patient. In doing this, we will follow the same procedure as in the previous section, starting with a consideration of how an agent may arrive at the belief in question and then arguing that the loss of sensory access to the patient also limits the possibility of forming this belief.

An agent's belief in the vulnerability and dependency of the patient was summarized as the belief in the possibility of acting, or neglecting to act, in ways that will affect the patient's physical or psychological health or well-being or the patient's sound exercise of cognitive and intellectual powers. Similar to the beliefs discussed in the previous paragraphs, the belief in the vulnerability and dependency of the patient may also *in part* be arrived at by means of evidence indicative of the particular state of the patient and by means of evidence indicative of the particular nature and properties of the setting or circumstances. This may, it seems, be the case in several ways. Thus the belief in the possibility of influencing the physical and psychological well-being of the patient and the sound exercise of cognitive and intellectual powers may *in part* be arrived at on the basis of evidence of properties of the patient or the setting indicative of the patient being causally affected by the agent in ways that either increase or decrease the physical or psychological well-being of the patient or the soundness of the patient's exercise of their cognitive capabilities. This could, to take the example of the previous paragraph, be evidence of a change in the patient's facial expression, such as the appearance of tears following an utterance of the agent. However, it seems as though the belief in question may *in part* also be arrived at on the basis of evidence of properties of the patient or the setting indicative of the patient already suffering from physical disability or of disabling perplexity, irresolution, ignorance, apathy, insecurity, fear, despair, depression, unfounded prejudice, lack of a capacity for reasoning and so on. Evidence of properties indicative of the actual suffering of the patient would generally form part of the grounds for holding the patient to be vulnerable in the specified sense simply because suffering is, it seems, already believed to be on a gradated scale but also something that may be influenced by human action. Thus, for instance, evidence of the patient crying or of a glass of pills on the table next to the patient may indicate some sort of suffering, which is then taken to reflect the vulnerability of the patient simply because it is already believed that suffering is remediable by human action. With this in mind let us one last time consider the conditions of gathering evidence in cyberspatial interaction.

At this point we may simply recapitulate the implications of the loss of sensory access to the patient and to the setting of the patient in cyberspatial interaction outlined at the end of the previous section. The loss of sensory access to the patient was here claimed to imply the loss of evidence of the patient's facial expressions, exclamations, tone of voice, gestures, other bodily movements, colour of skin, clothing and so on, all of which may be indicative of the emotional, psychological and intellectual state of the agent. Likewise the loss of sensory access to the setting of the patient implies the loss of evidence of the basic properties of entities constituting the patient's environment, such as size, shape, colour or smell, which may in turn be indicative of the properties and character of the entities constituting that environment. It should be clear that the loss of evidence of these properties of the patient and setting implies a loss of evidence relevant for the formation of the belief in the vulnerability and dependency of the patient.

Given that we arrive in part at a belief in the vulnerability and dependency of the patient as described above, it is clearly the case that the loss of evidence indicative of the state of the agent and of the character of the setting implies, other things being equal, the loss of evidence relevant for the formation of these beliefs.

Justification, evidence and cyberspatial interaction

In the previous three sections we tried to show that certain beliefs concerning a patient with whom one is interacting may be arrived at on the basis of evidence indicative of the properties of the particular patient or the setting. Therefore evidence indicative of these properties is relevant for the formation of the beliefs in question, and consequently the loss of this evidence in cyberspatial interaction is a loss of evidence relevant for the formation of the beliefs in question. However, it is one thing to say that an agent *may* arrive at these beliefs on the basis of evidence indicative of the properties of the patient or setting. For the loss of this kind of evidence in cyberspatial interaction to pose a problem would require that an agent *would* actually form the relevant beliefs in this way in the relevant situations: in other words, that the agent would form beliefs in the relevant way because of the kind of situation or a disposition of some sort, or that the agent would form the beliefs this way in order to justify them. In line with our considerations at the end of the preceding chapter, we will propose a combination of the two. Thus we will claim that we have a disposition to justify beliefs on the basis of observational evidence in cases where the beliefs may in any way be justified by reference to observational evidence – or, to put it slightly differently, that our everyday practices of justifying beliefs draws heavily on observational input, or simply on evidence.

First, it is worth noting some of the different ways in which we may try to justify a belief. It seems as though one may sometimes justify a belief on the basis of being in a certain mental or emotional state. Thus, for instance, we would sometimes justify our belief that another person is not reliable on the basis of having this particular 'intuition', which we may not be able to link further to the properties of the person or situation. In another case we may justify the belief that we have caused another person to be upset by taking the belief to be the result of divine inspiration. Third, we may justify our belief that the sum of two uneven numbers is always an even number on the basis of pure reflection on the general character of adding two numbers. In all of these cases the justification for the relevant belief is not based on evidence in the sense of input gained by means of our external senses – the justification simply does not involve seeing, hearing, tasting, smelling or touching in any ordinary sense – despite the fact that at least two of these beliefs concern something that may be taken to be partly justifiable by means of evidence. Although we may justify beliefs in this way, it seems as though there is a more commonly shared concept of justification according to which a belief concerning something that may be accessed directly or indirectly by means of our external senses or that may be inferred from what is accessed directly or indirectly by means of our external senses is justified to the extent there is evidence in support of the belief. That is, it seems as though we would generally tend to look for evidence in support of a belief that may be supported in this way – and we would require others to provide evidence of their beliefs that may be supported in this way. If someone believes that they have upset another person, for instance, we would, it seems, expect that person to provide some evidence of this being the case – for example, to interrupt the conversation and without further explanation get up and leave. The claim we are making here is simply that we generally consider observational evidence to carry significant weight in terms of justification, and (consequently) that whenever we are trying to justify a belief that may be justified by means of observational evidence we would be disposed to do so by means of observational evidence. This is an assumption that we will not attempt to ground further; instead we will move on to note the possible ways of justifying a belief by means of observational evidence.

Note that there may be more than one way of justifying a belief on the basis of evidence of the patient and the setting. Thus a belief in the determinateness of the patient may be formed by inference from the analogy between the situation in which the agent and patient are interacting and some other situation in which the agent interacted with another patient for whom the assumption of determinateness held.

In trying to justify the belief in the determinateness of the patient the agent may then try to build evidence of the similarities between the two situations – instead of building evidence of each of the properties of the patient as suggested earlier. However, as long as the belief in the determinateness of the patient is formed and justified on the basis of evidence of the patient and setting, then the consequences of any interaction in cyberspace for the availability of relevant evidence also seem to apply. That is, regardless of the specific way in which the observational evidence enters into the formation of justified beliefs regarding the patient, the decrease in the availability of relevant evidence would affect the possibility of forming justified beliefs about the patient in cyberspatial interaction.

Before closing this section, let us briefly consider what would happen if we were to give up the assumption made above. We have been trying to prove the point that normal humans have a disposition to justify beliefs that may be so justified by means of observational evidence. This means that the loss of evidence in cyberspatial interaction implies the inability of *any* agent engaging in this activity to justify those of their beliefs about the patient that could and otherwise would have been justified by means of observational evidence – and this in turn implies that the relevant beliefs will not be held and acted on by agents, since in the preceding chapter we argued that justification is of major importance for our goal of building, and acting on, knowledge. By rejecting the assumption that we generally seek to justify our relevant beliefs on the basis of observational evidence, the general conclusion must be dismissed. In other words, it may be considered possible to justify such beliefs without observational evidence of the patient and setting and thus possible to hold cardinal beliefs in the determinateness and vulnerability of the patient as well as beliefs in the character of certain causal relations without observational evidence about the patient and setting. The conclusion would still hold, however, for those agents trying to justify their beliefs on the basis of observational evidence about the patient and setting. Moreover, since we have in the three preceding paragraphs argued that an agent may arrive at a belief in the determinateness and vulnerability of the patient as well as at the belief in having a causal effect on the patient constitutive of the agent's own life-world on the basis of observational evidence, it would hold for any agent trying to build the relevant beliefs on observational evidence about the patient and setting that the interaction with the patient in cyberspace implies the loss of evidence relevant for the agent's formation of these cardinal beliefs.

7.1.3   Lack of reliable evidence

The preceding sections argued that cyberspace limits the availability of evidence relevant for the formation and justification of certain key beliefs regarding the patient. This argument is continued here through an investigation of the reliability of the information, and thus of the evidence, available in interaction in cyberspace: that is, an investigation into whether or not the available evidence is reliable as evidence indicative of the relevant properties of the patient and the setting. As will become clear, the third key property of interaction in cyberspace, namely the anonymity of the interacting parties, plays a key role in our considerations.

Experimenting with identity

Two things seem to point towards the unreliability of the evidence available in the relevant kinds of interaction in cyberspace.

The first is that the anonymity of interaction in cyberspace enables extensive experimentation with identity. Let us define experimentation with identity as an event involving a deliberate staging of identity in interaction on the basis of expressions that, when interpreted in accordance with their customary use, convey the equivalent of a false statement regarding the properties constituting the individual's personality, where the falseness of the statement implies the non-correspondence between the content of the statement and reality. So defined, such experimentation is clearly not restricted to cyberspace but may also be conducted in face-to-face interaction. There is a difference, however, in the extent to which a person may experiment with identity inside and outside cyberspace. Thus the anonymity of cyberspace clearly allows for fairly comprehensive experimentation with identity (compared with the possibilities outside cyberspace) in terms of properties such as age, gender, race, values, psychology and emotional state, personal history, aesthetic taste or financial situation. What we have shown here is only the conceptual possibility of extensive experimentation with identity in cyberspace in the sense of it providing the anonymity required to mislead people into forming a significant quantity of false beliefs regarding an agent's personality. We will, however, take our claims regarding the role of anonymity in interaction in cyberspace beyond the realm of the merely logically possible.[4]

---

[4] On the possibility of experimentation with identity in cyberspace see also Stanovsky in [91, pp. 174–175].

The second indication is that the possibility of experimenting with identity is occasionally – as a contingent, empirical matter of fact – taken advantage of by people interacting in cyberspace. On the basis of the available evidence, it appears to be a fact that people occasionally experiment with their identities in cyberspace in the sense that they deliberately adopt roles on the basis of expressions that, interpreted conventionally, convey the equivalent of a false proposition regarding the properties constituting their personality.[5] As has been said, we will leave aside any attempt at metaphysically grounding this claim, not simply because this would be beyond the reasonable scope of our present endeavours but also because this fact alone is all we need to pursue the argument of this section. By referring to people experimenting with their identity instead of just people involved in deception, however, we aim to signal that there is an explanation of what amounts to deception and that this explanation is rooted in the human nature, but also that the possible explanation of such natural human deception need not proceed along the Hobbesian line and thus be based on a postulate regarding human beings pursuing the gratification of desires with non-compliance to moral conventions being a possible result. (This is discussed in more detail below.)

Although we will not delve into the metaphysical background of the claim, it is worth noting that the claim is very modest and must therefore be distinguished from a similar but significantly stronger claim. The claim is weak in the sense that it does not claim a difference between the extent to which people experiment with their identity inside and outside cyberspace. That is, it does not claim that the extended anonymity of cyberspace may encourage more experimentation with identity in the sense that people would experiment with their identity more often. Although this may very well be the case, and the strong claim may therefore be justified, we will continue our line of reasoning on the basis of the weak claim. It has also to be noted that the modesty of the claim is easily recognized when people's everyday behaviour is taken into consideration. Thus it seems as though at least some people occasionally experiment with their identity in everyday situations in which they enjoy anonymity – perhaps by trying out opinions in discussions with strangers where such experimentation may sometimes be for the sake of argument – but at other times may also engage in an element of experimentation with identity in the sense referred to. Several other examples could be given, but if it is granted that such experimentation may be performed

---

[5] This fact has considerable evidence in the literature on interaction in cyberspace: e.g. Turkle in [97, pp. 177–209] and Rheingold in [83, pp. 149–180].

by ordinary people in ordinary everyday life outside cyberspace in situations in which they enjoy anonymity, then there seems to be good reason to suppose that it may also happen in interaction in cyberspace, where anonymity is also enjoyed. We will therefore assume the soundness of the contingent, empirical claim in its weak form and return to the main argument.

Hobbesian psychology

The unreliability of the evidence exchanged in cyberspace may also be grounded in more extensive considerations of the nature of human beings.

One such account of the nature and state of human beings is based on the following claims: (1) human beings are disposed to pursue the maximization of utility, where utility is defined as the satisfaction of preferences (desires);[6] (2) human beings are egoistic and egocentric in the sense of being wholly unconcerned about the satisfaction of the preferences of others;[7] (3) because of the scarcity of goods and because of their preferences,[8] human beings are constantly placed in situations of conflicting preferences – i.e. situations in which one person's maximization of utility is incompatible with another person's maximization of utility in the sense that their maximization of utility involves inflicting harm on others either directly by reducing the utility of another person or indirectly by removing goods that would otherwise have become sources of utility for the other; and (4) human beings would generally prefer situations in which they do not maximize utility by inflicting harm on others and in which others do not maximize utility by inflicting harm on them to situations in which they maximize utility by inflicting harm on others and others maximize utility by inflicting harm on them.[9]

In terms of the first claim it is worth noting that it is not saying that as humans we are disposed to maximize the satisfaction of preferences, but rather that as humans we seek to maximize the utility constituted by the satisfaction of preferences. This reflects the fact that we would not consider a state of the world satisfying one preference to be on a par with a state of the world satisfying another preference on the grounds

---

[6] Cf. Hobbes in [46, pp. 100–101].

[7] Cf. Hume in [48, sect. iii., part i.] and Hobbes in [46, pp. 100–101]. Cf. also Gauthier in [31, pp. 114–115].

[8] Cf. Hume in [48, sect. iii., part i.] and Hobbes in [46, pp. 100–101]. Cf. also Plato [76, p. 42 (Book II)], Gauthier in [31, pp. 114–115] and Rachels in [81, pp. 139–141].

[9] Glaucon in Plato [76, p. 42 (Book II)].

that they both satisfy a *single* preference. Rather, we would assign the equivalent of a numerical value to each of the states of the world at which the preferences apply, where this assignment of value would determine our choice of action – provided, of course, we would take the actions to have an equal probability of bringing about the relevant states of the world. It is this subjective value we refer to as 'utility'.[10] In terms of the second aspect it is worth noting the strength of the egoism and egocentrism introduced. In the definition given here egoism and egocentrism may not include having a preference for the satisfaction of the preferences of another person. Other definitions of egoism may be broader and hence include such preferences. In terms of the third and fourth claim it is worth elaborating slightly on the structure of the situations of conflicting interests. For the sake of simplicity let us, just for the present, suppose that the situation involves only two agents, say A and B, each having the choice between maximizing their utility by harming the other and simply refraining from doing so. Given their possible choices of action, we thus end up with a situation with four possible outcomes, which may be depicted as below, where the numbers indicate how the agents would rank the outcome $(O1 - O4)$ on the basis of the utility gained:

$$
\begin{array}{c c c}
A \backslash B & \text{Harm} & \neg\,\text{Harm} \\
\text{Harm} & 3,3\ (O1) & 1,4\ (O2) \\
\neg\,\text{Harm} & 4,1\ (O3) & 2,2\ (O4)
\end{array}
$$

As should be clear, both agents prefer the outcome where they maximize their utility without suffering any harm – and the worst possible outcome is to refrain from maximization but still suffer harm. In accordance with the fourth claim made above, both would prefer it if both do not maximize and hence do not suffer harm rather than both maximizing and hence suffering harm.[11]

Let us now consider how an agent disposed to maximize utility would act in the given situation. Let us assume that any agent disposed to maximize utility would act the way that they believe would maximize their utility, given their beliefs regarding circumstances and the possible choice of actions of others. That is, any agent would act in the way from which they would expect to benefit the most – in short, they would act so as to maximize their expected utility. In situations with the structure outlined,

---

[10] Cf. Gauthier in [31, p. 22].

[11] The structure of the situation outlined in the matrix is also known as a 'Prisoners' Dilemma' structure. Cf. Gauthier in [31, p. 79].

agents acting so as to maximize their utility independently of any indications and knowledge of how others will act would clearly act so as to harm the others simply because they may then expect, at best, to end up with the favoured outcome and, at worst, end up with the second to worst outcome, whereas agents may only expect, at best, to achieve the second to best outcome and, at worst, suffer the least preferred outcome if they were to refrain from doing harm to others. However, in acting with the consequence of doing harm to others they would all end up with an outcome second to worst ($O1$). Given this, it seems as though the agents would be better off if they made and acted in accordance with an agreement not to harm each other. In so doing they would all end up with the second to best outcome ($O4$). For this to be the choice of action for the maximizing agents it would, however, have to be shown that the agents would maximize their utility by actually complying with the agreement. Thus it seems as though agents would be even better off if they made an agreement not to harm others and if others acted in accordance with the agreement, but agents themselves violated it. Let us consider this objection to compliance with an agreement in further detail.

Suppose the utility gained by an agent if they both fail to comply the agreement is denoted u ($O1$), the utility gained by an agent if they both comply with the agreement is denoted u' ($O4$), and the utility gained by an agent by not complying with the agreement while the other complies is denoted u'' ($O2$ and $O3$), then it clearly holds that $u'' > u' > u$.[12] Suppose now that an agent disposed to comply with the agreement is defined as an agent who will comply with the agreement if others comply but who will not comply with it if others do not do so. Given the probability p of others acting in accordance with the agreement, the agent so disposed may expect an overall utility from the outcome of interaction to be calculated as: $[pu' + (1 - p)u]$. Suppose, further, that an agent disposed not to comply with the agreement is defined as an agent who will not comply with the agreement whether or not others comply. Again, given the probability p of others acting in accordance with the agreement the agent so disposed may, it seems, expect an overall utility from the outcome of her interaction of: $[pu'' + (1 - p)u]$. Since $u'' > u'$ then $[pu'' + (1 - p)u] \geq [pu' + (1 - p)u]$ – hence an agent disposed not to comply with an agreement may expect to reap more (or an equal amount for $p = 0$) utility from this disposition than an agent disposed to comply with an agreement. Consequently the objection against compliance with an

---

[12] Gauthier in [31, pp. 171–172]. See also for the following definitions and calculations.

agreement is sound. This reasoning, however, rests on a crucial assumption: namely, that agents disposed not to comply with the agreement are able to deceive other agents into believing that they will comply with the agreement. If deception is taken to be impossible, the calculation of the utility of the agent disposed not to comply with an agreement would yield a completely different result. Thus agents disposed to comply with the agreement would recognize agents so disposed and exclude them from the protection of harm implied by the agreement. That is, the agent disposed not to comply with the agreement may expect an overall utility from the outcome of interaction of u. Since $u' > u$, then $[pu' + (1 - p)u] \geq u$ – hence an agent disposed to comply with an agreement may expect to reap more (or an equal amount for $p = 0$) utility from this disposition than an agent disposed not to comply with an agreement. Consequently the objection against compliance with an agreement is not sound.

In our attempt to assess the reasonableness of the objection raised above, we have calculated the utility of agents disposed to comply or not to comply with an agreement not to harm each other under assumptions of either complete opacity or complete transparency of the dispositions of the agents. In order to make these considerations more in line with the workings of the real world, let us now assume agents to be translucent in the sense of being neither opaque nor transparent – in other words, let us assume that agents to some degree reveal properties indicative of their respective dispositions.[13]

The translucency of agents clearly introduces the possibility of agents failing to recognize the disposition of others. For the purpose of calculating the utility of the variously disposed agents interacting in conditions of translucency, we therefore have to introduce several different probabilities. Thus let p be the probability that agents disposed to comply with the agreement recognize each other, let q be the probability of these agents being recognized by those disposed not to comply without recognizing them, and finally, let r be the probability of a randomly selected agent being disposed to comply with an agreement. Let us further assign the utility of the best possible outcome for an agent ($O2$ for agent A) the value 1, the utility of the second to best outcome ($O4$) the value u', the utility of the second to worst ($O1$) the value u, and finally the worst outcome ($O3$ for agent A) the value 0. This means that $1 > u' > u > 0$. We may now calculate the overall utility the differently disposed agents may expect from interacting with other agents. First, any agent disposed to comply with

---

[13] Gauthier in [31, p. 174].

an agreement would clearly be able to expect an utility of u from interacting with others if they were neither: (1) to enter an agreement with others equally disposed; nor (2) to be exploited by somebody disposed not to comply. The probability of (1) is the combination of the probability of the other being equally disposed (r) and of the two recognizing each other (p): i.e., rp. The chance of entering an agreement obviously increases the agent's expected utility, and this *increase* is given as $[rp(u' - u)]$. The probability of (2) is the combination of the probability of the other being disposed to exploit $((1 - r))$ and of only the other recognizing the disposition of the agent (q): i.e., $(1 - r)q$. The chance of being exploited clearly decreases the agent's expected utility, and this *decrease* is given as $[(1 - r)q(u' - 0)]$. The overall utility that the agent disposed to comply with an agreement may expect is thus given as:

$$u + [rp(u' - u)] - [(1 - r)q(u' - 0)]$$

Second, any agent disposed not to comply may expect an overall utility from interacting with others of u if they do not exploit someone disposed to comply. The probability of exploiting an agent disposed to comply is the combination of the probability of the other being disposed to comply (r) and of any agent recognizing the disposition of the other without themselves being recognized (q): i.e., rq. The chance of exploiting someone disposed to comply obviously increases the agent's expected utility, and this *increase* is given as $[rq(1 - u)]$. The overall utility that an agent who is disposed not to comply with an agreement may expect is thus given as:

$$u + [rq(1 - u)]$$

In the light of these calculation we may now return to the question of whether or not any agent may expect to maximize their utility by complying with an agreement not to harm others. For this to be the case it would have to hold that the expected utility from compliance would be greater than the expected utility from non-compliance. Interacting in conditions of translucency this would be the case if, and only if:

$$u + [rp(u' - u)] - [(1 - r)q(u' - 0)] > u + [rq(1 - u)]$$

This is the equivalent of:

$$p/q > [((1 - u)/(u' - u)) + ((1 - r)u/r(u' - u))]^{14}$$

---

[14] Cf. Ploug in [79, p. 212].

Thus an agent may expect to maximize utility by complying with an agreement not to harm others if, and only if, the ratio between the probability of the interaction resulting in co-operation between agents disposed to comply and the probability of the interaction resulting in exploitation is greater than the ratio between the gain from non-compliance and the gain from co-operation (since $((1-r)u/r(u'-u))$ is always equal or greater than 0).[15] Since any agent's prospect of maximizing their utility thus comes to depend on the ratio between the probabilities p and q, it is clearly of interest to know how an agent may affect these. The probability p depends on the ability of compliant agents to discover the same disposition in others as well as the ability to reveal their own disposition to them. The probability q depends on the ability of compliant agents to discover non-compliant agents and to hide their disposition from these, but also on the ability of the non-compliant agents to discover the compliant agents and to hide their own disposition from them.[16] It is worth noting how changes in the ability to recognize the dispositions of other agents more specifically affect the prospects of maximizing utility. If agents generally improve their ability to recognize the dispositions of others, the value of p will obviously increase, but at the same time q will stay relatively constant since the increase in the ability of the non-compliant agents to recognize the compliant agents is countered by an increase in the ability of the compliant agents to discover non-compliant agents (given more or less the same increase in the ability). Consequently the value of p/q will increase with the implication that other conditions (notably r) have to be less favourable in order for compliance with an agreement to be a way of maximizing expected utility.[17] However, if on the other hand agents become less able to recognize the dispositions of others the value of p will decrease, and once again the value of q will stay relatively constant (given more or less the same decrease in the ability). Consequently the value of p/q will decrease – and approximate to 0 – with the implication that agents may not expect to maximize their utility by complying with an agreement not to harm others. Actually an agent may not expect to maximize utility from the point when $q \geqq p$ as $((1-u)/(u'-u)) \geqq 1$ as $1 > u' > u$ and $((1-r)u/r(u'-u)) \geqq 0$). This result seems to hold for any situation in which it is at all possible to recognize the disposition of the other agents; i.e., for any situation in which $p > 0$.

---

[15] Cf. Gauthier in [31, p. 176].

[16] Cf. Gauthier in [31, p. 180].

[17] Cf. Gauthier in [31, p. 181].

Given these considerations, it clearly becomes relevant to consider what may and may not improve an agent's ability to recognize the dispositions of others. It seems as though possible ways of recognizing the disposition of other agents to comply with an agreement not to harm each other are ways of recognizing their commitment to compliance, where lack of commitment may be recognized by virtue of an agent showing signs of dishonesty and indifference towards the agreement, or simply by virtue of the agent being revealed as someone who formerly reneged on an agreement. Granted that these are possible ways of recognizing the disposition of others, it is clearly the case that recognition of the disposition of others requires evidence indicative of the properties of the agents. In assessing whether or not somebody will comply with an agreement an agent has to have some direct or indirect evidence indicative of the properties of that other person. Consequently, when defined as the state of an agent in which the set of properties constituting the agent's identity is unknown to others,[18] anonymity is, other things being equal, a threat to both the improvement and the sound exercise of this ability. More specifically, the extensive anonymity of agents interacting in cyberspace, or simply the lack of sensory access to others in cyberspace, is clearly a threat both to the development of the ability to recognize the relevant dispositions of others and to the sound exercise of this ability. As we have already argued, the limited sensory access to other agents in cyberspace limits the availability of evidence indicative of the properties of agents and thus also of properties that may be indicative of a commitment to an agreement not to harm each other. However, the limited availability of evidence indicative of the properties of an agent also implies limited possibility of even recognizing agents, and thus one of the most important ways of recognizing the disposition of other agents – recognition based on former experience – is made impossible. As an illustration of the extensive difficulty in recognizing other agents in cyberspatial interaction it is worth highlighting one of the features of chat-rooms. This is that in chat-rooms the interacting people know each other only on the basis of a self-selected nickname, which may be changed at any point in time. The only evidence of the properties of another agent available in this kind of interaction that may serve the purpose of recognizing the agent and the disposition of the agent is, therefore, the content as well as other features of the written message.

Although information is exchanged in cyberspatial interaction, the amount that may serve as evidence indicative of the properties of agents and their dispositions

---

[18] Cf. the definition of anonymity in Chapter 4.

is very limited. Hence the probability of recognizing the dispositions of anonymous others seems to be very low. This means that the probability p of compliant agents recognizing other compliant agents is very low, and agents may thus not expect to maximize their utility by complying with an agreement not to harm others. Expressed slightly differently, it seems as though agents disposed to maximize their utility would do better in cyberspatial interaction if they did not comply with an agreement not to exploit others. Note, importantly, that if agents act on the basis of this reasoning, then they may actually further decrease the probability of recognizing the dispositions of others by lying about their own disposition. Hence the reliability of one of the only sources of evidence indicative of the agent's disposition in chat-rooms – the content of the written messages – disappears.

The reliability of evidence

In our previous analysis of the notion of reliable evidence we concluded that the reliability of evidence is dependent, among other things, on the beliefs of an agent such that a change in an agent's beliefs may change what the agent conceives to be reliable evidence. It is the aim of this section to show that experimentation with identity in cyberspace and the possibility of others acting so as to maximize their utility in cyberspatial interaction may influence the beliefs of agents interacting in cyberspace to change their view about the reliability of the evidence gathered there. As should be clear, the presupposition of establishing such a link between beliefs regarding the reliability of evidence and the actual experimentation with identity or the behaviour of others directed at maximizing utility is that agents interacting in cyberspace do in fact form beliefs about these actual and possible actions of others. We will therefore support our attempt to establish such professed links by considering whether an agent may be at all aware of these actual and possible actions of others.

As regards the general belief in or awareness of it being the case that agents experiment with their identity in cyberspace, this may be argued to be an extrapolation from all agents' own experience of experimenting with their identity. As already argued in a previous section, it seems plausible that we sometimes and for a variety of reasons such as curiosity or insecurity, experiment with our identity when we enjoy some kind of anonymity, for example by trying out political opinions, religious convictions or ethical values in everyday conversations with strangers. Since there seems to be no natural and striking reason to suppose that we experiment with our identity because we are in principle different from other people, it seems as though the belief in

and awareness of our actual and possible experimentation with our identity are really
a belief in and awareness of it being the case that human nature encompasses a potential for experimentation with the constituents of their identity. In short, it seems as
though the belief and awareness that we ourselves experiment with our identity when
enjoying anonymity is naturally extended to a belief and awareness of others similarly
experimenting with the constituents of their identity. If it is granted – perhaps on the
grounds just suggested – that we generally believe in or are aware of the possibility of others experimenting with their identity in interactions in which they enjoy
anonymity, then we have taken the first step towards substantiating the fact that in
cyberspatial interaction we cannot rely on the evidence indicative of certain of the
patient's properties to the same degree as we would in face-to-face interaction.

The second step is to argue that deciding whether or not a person is experimenting
with their personality or identity is difficult in those kinds of cyberspatial interaction
in which the anonymity of a person is not threatened. One possible way of deciding
whether or not another person really is the person they are pretending to be is to
collect evidence indicative of properties that we believe a person cannot consistently
possess at the very same time. For example, if someone is claiming in an extremely
low-pitched voice to be a woman, then we may under certain circumstances take
this to indicate that the person is lying. In general, however, inconsistencies between
statements may not be a very reliable way of deciding the trustworthiness of the
evidence indicative of the properties of another person. Thus they may reflect fallacies
and imperfections of both the source and receiver of the statement, while consistency
between statements may reflect the integrity of their source. Moreover, inconsistencies
between different kinds of evidence are hard to establish in interaction in cyberspace.
The example of an inconsistency given above may be considered to be an inconsistency
between different kinds of evidence indicative of the properties of another person,
namely the evidence provided by the content of the statement 'I am a woman' and
the evidence provided by the pitch of the voice. In interaction in chat-rooms such
exact inconsistency is impossible to establish, simply because the evidence provided
is only in the form of written statements. Hence not only may inconsistencies between
statements not be reliable as a way of disclosing unreliable evidence but the availability
of kinds of evidence that may serve the purpose of revealing inconsistencies is limited.
At the same time there may be consistency without truth.

Given that we believe others may be experimenting with their identity in our
cyberspatial interaction with them, and given that we do not have very reliable ways

of determining this, it seems as though the conclusion is unavoidable that we cannot
rely on the evidence indicative of certain of the patient's properties to the same degree
as we would in face-to-face interaction. That is, we cannot rely to the same degree
on the evidence indicative of those of the patient's properties of which we would
have evidence in increased quantity and variety in face-to-face interaction. This is a
significant result since anonymity in cyberspace extends to a host of properties of a
person and of identity that are not covered by the anonymity enjoyed in face-to-face
interaction. In text-based chat-rooms a patient may conceal gender, age, shape and
size of body, handicaps and so on. None of these properties may be concealed – at
least not to the same degree – in face-to-face interaction. In other words, if an agent
collects evidence indicative of the properties of a patient and of those properties
constituting the identity of the patient, the availability of reliable evidence is less
than in face-to-face interaction.

It may, of course, be objected that evidence is never wholly reliable – not even the
evidence collected in face-to-face interaction. If we leave aside the global epistemological scepticism which claims that there is no such thing as reliable evidence, then this
objection is really beside the point. It may very well be that there is always only a
certain probability of some evidence being reliable simply because of the ways of the
world and of the human beings in it: in other words, owing to human fallibility and
to limitations in reasoning and perception and so on. This, however, does not obviate
the point made above. The point advanced here was that, in assessing the reliability of the evidence of certain properties that constitute a personality, there exists a
difference between interaction in text-based chat-rooms and interaction face-to-face.
Further reflection on the nature of the human epistemological apparatus and *modus
operandi* may add to the general unreliability of evidence, but that unreliability will
then apply equally to evidence collected inside and outside cyberspace. With these
remarks in mind, let us move on to consider how the possibility of others acting so as
to maximize their utility may affect our belief in the reliability of the evidence gained
in cyberspatial interaction.

Once again the starting-point for these considerations is the attempt to establish
that we are generally aware of or believe in the possibility of others acting so as to
maximize their utility, where this may involve doing harm to or exploiting others, in
conditions in which they enjoy anonymity. There seem to be at least two strategies
in the attempt to support this claim. One is to assume that human beings are actually equipped with a Hobbesian psychology that makes them constantly engage in

evaluations and calculations of, for example, the benefit deriving to them from a given
chance to exploit others, and then to argue that they would engage in cyberspatial
interaction on the basis of such calculations and, therefore, that they would expect
everyone to maximize their utility by exploiting others. The other is to assume that
human beings are only partly equipped with a Hobbesian psychology, which may
occasionally make them act so as to maximize their behaviour by exploiting others
in conditions of anonymity, and then to argue that it is the threat of others being catapulted, so to speak, into action on the basis of their Hobbesian psychology
in conditions of anonymity that make us constantly aware of or to believe in the
possibility of others exploiting us in conditions of anonymity. We will here assume
the second strategy simply because Hobbesian psychology in itself seems inadequate.
For one thing, it seems as though people may sometimes – even under conditions of
anonymity – deliberately act so as to satisfy (a preference for the satisfaction of) a
preference of another being, where this will, on balance, result in achieving less utility
than they might otherwise have achieved (perhaps even a loss).

Imagine, for example, two people engaged in conversation in a chat-room. The conversation is very sincere and touches on a number of issues that are deeply emotional
for at least one of the two agents. Suppose now that the other person is suddenly
struck by a substantial lack of interest in the very emotionally involved person. At
this point the uninterested person may choose to leave immediately or may choose
over a period of time to end the conversation. It seems plausible to suppose that
immediately leaving may harm the emotionally involved person, but that the utility
the other has gained from a sincere conversation may be preserved by ending the
conversation in an appropriate manner. Given the anonymity afforded by interaction
in a chat-room, it seems as though the person has the choice between, on the one
hand, maximizing utility by satisfying a preference for a sincere conversation and a
preference for 'getting on with life' and, on the other hand, abstaining from maximizing utility by acting so as to preserve the utility of the other. If both these choices
are possible continuations of the scenario, then it seems as though we have shown the
reasonableness of agents occasionally acting so as to take advantage of the anonymity
afforded by cyberspace in the maximizing of their utility, but also acting so as to
consider the utility of others under conditions of anonymity.

Returning to the main line of argument, what has to be shown now is that it is
likely that we are aware of or believe in the possibility of others simply exploiting us.
The argument is simply this: since we are ourselves partly equipped with a Hobbesian

psychology, it seems as though we are, when acting under conditions of anonymity, constantly confronted with the possibility of exploiting others, and since there seems to be no striking reason to suppose that we are in this respect any different from other people, we may infer that they are also in conditions of anonymity faced with the possibility of acting so as to exploit others. If it can be granted – perhaps on the grounds just suggested – that generally we believe in or are aware of the possibility of others acting so as to maximize their utility by exploiting others in conditions of anonymity, then we have in reality given a strong argument for the evidence provided by others being considered less reliable in conditions of anonymity than the same evidence provided in face-to-face interaction. Thus if agents are to maximize their utility by exploiting others in conditions of anonymity, then they obviously have to provide evidence indicative of who they are that does not threaten the anonymity enabling the exploitation of others. That is, in the event that agents take advantage of their anonymity to exploit others, they may have to provide unreliable evidence of who they are. In a chat-room this means that the interacting agents may provide written messages in which they convey false statements regarding their age, gender, occupation, values, location, income, political and religious affiliations, ethnicity, looks, physical and psychological health, education, criminal record, sexuality, emotions, intentions, disposition and so on. Hence, given that agents interacting in a chat-room are aware of the possibility of others acting with the intention of exploitation, they also seem to have reason to believe that the others may provide unreliable evidence of certain characteristics that those individuals would not have been able to provide in face-to-face interaction. That is, they seem to have reason to believe that the evidence provided is less reliable than the evidence provided in face-to-face interaction.

We may thus conclude that insofar as agents may be experimenting with their identity or maximizing their utility by exploiting others when interacting in conditions of anonymity, and insofar as these are possibilities we are generally aware of and take into consideration – perhaps in the ways suggested – then the evidence exchanged in, for example, chat-rooms in cyberspace may be considered to be, in general, less reliable than evidence exchanged in face-to-face interaction. That is, an agent interacting in a chat-room in cyberspace has reason to believe that the evidence exchanged is not as reliable as the evidence exchanged in interaction outside cyberspace.

Chapter 8

Belief and action

In the three previous chapters we argued that an agent's conviction concerning the reality of a particular patient is tied to the formation and justification of certain other beliefs, and that the transition from interacting outside to interacting inside cyberspace implies the loss of relevant and reliable evidence. Hence the conditions enabling the formation and justification of the belief in the reality of a particular patient are less favourable in cyberspace than outside cyberspace, and an agent may consequently come to be less convinced of the reality of a particular patient.

For this result to be relevant for our attempt to explain the assumed difference in interaction inside and outside cyberspace, it has to be shown that the belief in the reality of a particular patient is relevant for the way in which an agent may come to act. That is, it has to be shown that agent A's belief that p is a prerequisite of A's motivation to do q, and so of A doing q. In doing that, we will have covered the third stage of our attempt to explain the basic premise (see the first section of Chapter 5).

In the fourth stage the strings of the three preceding stages are pulled together in making it explicit that agent A may be less convinced of – or simply lack – the belief that p in interaction in cyberspace and therefore may fail to do q. In effect the fourth stage then simply provides an explanation of the relationship captured in basic premise.

In this chapter both the third and the fourth stage is covered. The chapter therefore marks the end of the attempt to explain the basic assumption taking into account the role of another person's face and our sense of reality.

T. Ploug, *Ethics in Cyberspace: How Cyberspace May Influence Interpersonal Interaction,*   **187**
© Springer Science+Business Media B.V. 2009

8.1   Belief, reality and ethics

8.1.1   Belief, reality and motivation

Showing that the belief in the reality of a particular patient may come to influence the ethical character of an agent's behaviour is a requirement that has at least two components. First, we need to show that beliefs may influence an agent's motivation; and second, we need to show that the specific belief in the reality of a particular patient may influence the behaviour of the agent with respect to its ethical character.

We have already addressed the former extensively. In Chapter 3 we saw how an agent's background beliefs may influence what, in a Humean view of moral motivation, the agent may consider to be an appropriate means to an end – or what, in a non-Humean view of moral motivation, an agent believes to be a reason for action and, consequently, what he will be motivated to do. This general result does not, however, entail any given specific belief belonging to the set of background beliefs that in one, more or perhaps even all situations influence behaviour in such a way that a change in the relevant belief may lead to a change in behaviour with respect to its ethical character. Thus in order for us to show that an agent's belief in the reality of a particular patient may influence the agent's behaviour in an ethically relevant way we need to provide an argument to the effect that the belief in the reality of the particular patient is one of those background beliefs that in one, more or perhaps all situations influence what an agent is motivated to do.

In order to show this we will make a rather trivial observation: namely, that when an agent comes to be motivated to extend moral concern to a particular patient on the basis of beliefs concerning the relevant properties of the patient, this motivation is preconditioned by a belief in the reality of the particular patient towards whom the moral concern is directed. In other words, it is preconditioned by a belief in the reality of the patient as a person with the relevant set of properties attributed to him or her by the agent. Put slightly differently, for any case in which normal human beings act to extend moral concern to a patient on the basis of beliefs concerning the nature and state of the patient it seems to hold that such action is preconditioned by a belief in the patient possessing the attributes relevant for extending that moral concern. Consequently no normal human being would extend moral concern to a patient if they came to believe that the patient did not possess properties that had been attributed to the patient and were relevant for extending moral concern.

Applied to a Humean conception of moral motivation the observation means that when an agent in situation S desires q and believes an action p to be the appropriate means to the desired end then it is the equivalent of the agent in situation S desiring q and believing action p to be the appropriate means to the desired end on condition that the entities constituting the situation are real in relevant respects. Thus, for example, any normal person having a desire to help a starving child in a photograph presented by a charity and holding the belief that the particular child may be helped by supporting the charity would only hold this desire and belief, and thus be motivated to support the charity, as long as the particular child was believed to be real and suffering from malnutrition, lack of shelter and so on: in other words, on condition that the child with the claimed properties was not the invention of some marketing strategist. The implication is that an agent coming to believe that the child on the poster is not really suffering would, other things being equal, neither desire to aid that specific child nor believe that the child may be helped by supporting the charity – and hence there would be no motivation to aid that specific child.

Applied to a non-Humean view of moral motivation, the observation means that when an agent believes there to be a normative reason to do p in a situation S then it is the equivalent of believing there to be a normative reason to do p in a situation S on condition that the entities constituting the situation are real in relevant respects. Thus, to use the example above, any normal person holding the belief that it would be morally right to help the starving child staring out from a photograph by supporting the representative of the charity holding the photograph would only hold this belief as long as the child was believed to be real as a child suffering from malnutrition, lack of shelter and so on: in other words, on condition that the child with the claimed properties was not the invention of some marketing strategist. The implication is that any agent coming to believe that the child is not really suffering will, other things being equal, not believe they have a moral reason, and hence will not be motivated, to aid that specific child.

Whichever way we construe moral motivation, we come to the same conclusion. To be motivated to extend moral concern to a particular patient presupposes that the particular patient towards whom the concern is extended is real in relevant respects.

It is worth noting the use of the notion of relevance in these considerations of the role of belief in the reality of the particular patient for an agent's behaviour. The dependency we have established between the motivation to extend moral concern and the belief in the reality of the particular patient is precisely between motivation and the belief in the reality of the patient in relevant respects. How does this more specific link between an agent's behaviour and the belief in the reality of the patient relate to the lack of evidence in cyberspace to support the beliefs that we previously argued underlie a more general conviction concerning the reality of a particular patient? The simple answer is that the strength of the specific belief is derived from the strength of the general belief. Thus if an agent engaged in interaction in cyberspace comes to be less convinced of the reality of the patient – i.e., less convinced that the patient embodies a certain set of properties – there is, other things being equal, no reason to suspect that the agent will hold particularly strong beliefs concerning the reality of the apparent morally relevant properties of the patient. That is, there is no reason to suspect that the agent will be convinced that the morally relevant properties of the patient are real and the morally irrelevant properties not so.

These considerations also raise the question of how moral requirements tie into particular situations. Before addressing this question, we will once again highlight the important conclusion in this subsection: namely, that our belief in the reality of the particular entities constituting a situation, including the reality of the patient, is a background belief, a change of which may lead to a change in behaviour.

8.1.2   The particularity of moral concern

As has already been suggested, the observation of the belief in the reality of a particular patient being assumed in any act of extending moral concern seems to presuppose that the particularities of a patient and the situation as a whole have some relevance for moral agency. We might then ask whether moral concern is not general in nature, in the sense that it is extended in response to rather abstract properties of the patient. That is, one may ask whether moral concern is not general in the sense that it is extended to other people on the basis of simply believing them to be human beings.

The question raised may be addressed as two separate questions. First, there is a question concerning the nature of moral requirements, principles and reasons. The question here is whether there are moral requirements, principles and reasons that are tied only to abstract properties of situations. Second, there is the question of whether our moral practice reflects a sensitivity to the particularities of situations and patients. Although it is clearly the second of these questions that is of interest for our attempt to explain the basic premise, we will look briefly at both of them.

The first question may be approached in a variety of ways. We will deal with it here by looking at how some of the main positions in normative ethics relate moral requirements to the properties of situations and patients. In Chapter 2 we briefly introduced three major theories in normative ethics: hedonistic utilitarianism, Kantian duty ethics and Aristotelian virtue ethics. Hedonistic utilitarianism claims that the morally right action is the one that maximizes sensual pleasure and minimizes pain for the greatest number of people under the given circumstances.[1] It follows that determining which action is the right one involves 'calculating' the possible outcomes of relevant actions – their probability and their contribution to the pleasure of the people affected by it – which in turn involves making assessments concerning not only what causes pleasure and diminishes pain, but also to what extent it causes pleasure and diminishes pain in the particular people involved. The resulting moral requirement clearly takes into account the people affected by an action not only as abstract human beings but also as particular individuals.

At the centre of Kantian duty ethics is the 'categorical imperative'. In one of its formulations, known as the 'formula of the end in itself', this states that one must act so as to treat humanity in others and oneself always as an end and never as a means only.[2] In Kant's own account, to treat others as a means is to perform actions that interfere with the lives of other people not only in ways they do not want or to which they do not consent but also in ways that do not provide them with the opportunity to consent: e.g., deception.[3] Although this may be interpreted as grounding a purely negative obligation, the Kantian requirement of non-instrumentalization has often been associated with the minimal positive obligation of obtaining informed consent before acting in ways that will affect another person. If such a positive obligation is assumed, it becomes clear how a Kantian ethics of duty must also take into account the particularities of a situation. Thus, if someone is to have consent to actions affecting others, then they will have to provide information to others on the planned action to varying degrees and in varying languages simply because they are to varying degrees able to understand and reflect on the information given. Again, this brief exploration into Kantian duty ethics seems to make it clear that the moral requirements in this theory incorporate others not as merely abstract human beings but as particular individuals.

If the moral requirements of a situation are determined by Aristotelian virtue ethics, then the relevant properties of an action are those that, taken together,

---

[1] Cf. Bentham in [5, pp. 14–15].

[2] Kant in [52, p. 79]. See also Scruton in [84, p. 86].

[3] Kant in [52, pp. 80–81ff.].

determine whether or not the action is a mean between excessive actions: i.e., between doing too much and doing too little.[4] What counts as a mean is dependent on the particularities of a situation. That is, the right action is tied to the situation in being dependent on the particular properties of the individuals being affected by the action and of other properties of the situation.[5] Aristotle talks of being good as the difficult exercise of addressing the right person in the right measure at the right time in the right way and with the right motives, where this may imply apparently conflicting ethical courses of action in different situations.[6] Even Aristotelian virtue ethics, therefore, ties moral requirements into the particular properties of individuals and situations – moral requirements do not concern other people merely as human beings but as particular human beings.

This brief exploration into three major theories in normative ethics clearly shows that it is a common trait of these theories that moral requirements are tied into the particularities of individuals. The moral requirement towards other people is not something that takes these other people into account only as human beings. Although the exploration is not exhaustive of ethical positions, it does serve the purpose of showing how three very different positions within normative ethics agree on the issue addressed here. Insofar as we subscribe to one of these positions, the resulting moral practice will be one in which the aim is to be sensitive to the particularities of situations in our moral choices. To put it slightly differently, if any one of these three theories comes to form the basis of our moral practice, then our moral practice will reflect a sensitivity to the particularities of other persons and situations. As stated at the outset of this section, our moral practice does not necessarily fit as neatly with a theoretical position as the ones outlined here. We therefore have to face the second question, of whether or not our moral practice reflects a sensitivity to the particularities of situations and patients.

The second question concerns our moral practice, by which is meant the way we actually interpret moral requirements in everyday life. Looking at our moral practice, it seems as though in many of our morally weighted actions we are responding to the particularities of other people. Most people seem to recognize a moral requirement to be polite. When applied in a given situation, it seems as though this requirement is thought of as incorporating not only the plain fact of another person being human but also the particularities of this person. Thus it seems as though most people – when

---

[4] Cf. Aristotle in [3, 1106a and 1106b].

[5] Cf. Aristotle in [3, 1106b, 1107a–1107b, 1109a–1109b].

[6] Cf. Aristotle in [3, 1109a–1109b].

applying the commonly accepted requirement of being polite in a given situation – would in varying degrees take into account such properties of people affected by their action as age, gender, religion, nationality, physical appearance and so on. On some occasions, therefore, we would think of the requirement to be polite as a requirement to greet another person by shaking hands instead of just saying 'Hello' simply because this other person is an older person. In a similar vein we may come to think of the requirement to be polite – or the requirement to do good – as a requirement to give up our seat on a bus to another person simply because this person is old, or to leave our position at the front of the queue in the post office to another person simply because this other person is a mother with a crying child. And yet again we may come to think of the requirement to be polite as a requirement to abstain from swearing in conversation with another person simply because that person considers certain swear words unacceptable for religious reasons. All of these examples seem to support the claim that in our everyday moral practice we respond in our understanding of the moral requirements in a given situation to the particularities of the people affected by our actions. We do not, it seems, incorporate others into our understanding of the moral requirements of the situation merely as human beings.

The claim made here may be countered on the basis of another example. For instance, when driving on the motorway most people would occasionally think themselves morally required to comply with the speed limit because of the dangers to which they would otherwise expose other people. We may be tempted to interpret this example as showing that we occasionally consider ourselves to be under a moral requirement in which other people are incorporated merely as human beings. This interpretation is, however, highly questionable. It seems as though the requirement to comply with the speed limit in order to avoid exposing other people to risk is precisely a requirement we would not act on in itself but apply to a given situation. Consequently, if we comply with the speed limit – or perhaps go slower than suggested by the speed limit – this reflects the application of the moral requirement to the given situation. That is, it reflects a consideration of how the moral requirement applies to the level of traffic, the weather conditions, the time of day and so on. Understanding oneself to be under a moral requirement to comply with – or to go slower than – the speed limit is thus the result of taking into account the particularities of the situation.

It might still be claimed that the example shows that the particularities of the individuals affected are not taken into account in the consideration of the moral requirement – the affected people are incorporated merely as human beings. The answer to this is threefold. First, in the relevant example the moral agent is constrained in taking into account the particularities of the other people affected by any moral choice by obvious epistemological limitations constituted by the drivers being in different cars. This does not show, however, that we do not strive to take into account the particularities of other people in our moral considerations to the widest possible extent – and that we think of moral requirements as partly constituted by the particularities of situations. Driving on the motorway thus differs from interaction in a chat-room precisely because it makes it physically impossible to obtain evidence of more specific and relevant properties of the individual. As elaborated on in a previous chapter, in a chat-room the epistemological limitations are also constituted by the physical conditions, but in a chat-room there is the additional problem of the reliability of the evidence received. The latter is an epistemological condition that may be influenced in the process of interaction. Through continued communication the reliability of the evidence of the other person may be tested – and this in turn may lead to the exchange of what is considered to be relevant and reliable evidence of the properties of the other person.

Second, it may also be claimed that the example captures one of those borderline cases in which no knowledge (or evidence) of the particular properties of the individuals affected by one's action is required in order to specify the applicable moral requirement. Assuming that most people would adhere to a principle of not doing harm to others, the claim here is that the common-sense interpretation of the implication of this principle is that exposing others for no good reason to an increased risk of accident, and hence of death or severe injuries, is, other things being equal, to do harm to others regardless of their age, gender, race, physical appearance, religion or other properties. Hence in this case the knowledge (or evidence) of others as human beings will do. But this is a borderline case: it marks the limit of the knowledge required of other people in order to understand the practical implications of the moral requirement of not doing harm to others. As should be clear, exactly the same principle will in many other cases require more knowledge (or evidence) of the particular properties of another person. If, for instance, this principle is thought to be relevant for situations in which an utterance may cause offence, then clearly the implications of the principle as a justification for a non-offensive behaviour will be dependent on particular properties of the individual, such as their political or religious convictions or their customs. To summarize: although our moral practice may at times seem to take others into account only as members of the class of human beings, these are borderline cases. In general, common-sense moral requirements, such as those of being friendly, polite, hospitable, loving, caring and so on, require an insight into the particularities of other people.

Finally, the example may also be claimed to be yet another instance of our failure to take into account the full spectrum of particularities relevant for an understanding of the moral requirement in a given situation. In itself the failure to take into account certain relevant particularities of deciding one's speed on a motorway may reflect a more general feature of moral life: namely, the inability to persist in the endeavour to make (fully) informed moral judgements. In other words, the failure reflects our incapacity to make moral judgements – the inability fully to dedicate our cognitive faculties to patiently and thoroughly deciding the morally right course of action. This may perhaps not be the most plausible interpretation of the example of determining our speed on a motorway, but it may be plausible as an interpretation of other situations in which we fail to take into account the particularities of a situation and the people affected by our actions and decide the moral requirement with reference only to the humanity of the people involved.

In response to the example above, in which it seems as though on certain occasions we understand the moral requirements of a situation without reference to the particularities of the people involved, we have suggested three interpretations of the example: that it reflects epistemological limitations rather than our understanding of the nature of moral requirements; that it reflects a borderline case in which the particularities of the people affected are not relevant for the moral requirement; and finally, that the failure to take into account the particularities of the people affected by our action reflects a more general inability in our moral psychology to be sensitive to the particularities of situations and people. Whether or not these are viable responses – on their own or in combination – may need further investigation and analysis. For our present purposes it seems to suffice, however, that we have shown how major positions in normative ethics construe moral requirements as partly constituted by the particularities of situations and the people in them, and that we have shown how, even when it has a broad focus, our moral practice takes into account the particularities of situations and the people in them. Together these results underpin the point we are trying to make concerning interaction in cyberspace: namely, that in interaction in, for example, a chat-room – as in life in general – we will constantly be faced with moral requirements corresponding to our own moral beliefs in being polite, friendly, helpful etc., and these moral requirements will partly be constituted by the particularities of the person to whom they are directed.

With this important result in mind we have finally laid the complete grounds for explaining the assumed difference between interaction inside and outside cyberspace. Before entering the fourth stage of our enterprise, one small note must be made. In this section we have more or less explicitly construed the situation of understanding a moral requirement as that of deliberately applying a principle or rule to a situation. This picture may be challenged. Aristotelian virtue ethics describes the situation of the moral agent understanding a moral requirement by means of a 'seeing metaphor' – the morally wise person 'sees' the moral requirement in a given situation.[7] Like other ways of construing the act of understanding a moral requirement, this does not affect the conclusion drawn. The morally wise person in Aristotelian virtue ethics is thus precisely as sensitive to the particularities of the situations as we have claimed.

---

[7] Cf. Aristotle in [3, 1113a].

8.2   Explaining the moral difference in interaction

We now enter the very last stage in our long journey towards providing an explanation of the basic premise on the basis of our inspiration from the Levinasian notion of the 'the face of the other' and from the 'Legal Tender' experiment. In the preceding chapters extensive investigations of our ways of forming certain beliefs and of how these beliefs underlie action have prepared the ground. Now it is time to pull all of the threads together in an explanation of the basic premise ([TBP]). This is done in three stages. First, we will summarize those results of the preceding chapters and sections relevant for an explanation of the basic premise. Second, we will provide an explanation of how an agent may fail to be motivated to act on a moral requirement in cyberspace. Third, we will consider to what extent we have accomplished the goal of providing an explanation of the basic premise. Let us start, however, by restating the basic premise:

[**TBP**]   Interaction in cyberspace occasionally differs from face-to-face interaction by virtue of being constituted by agents performing acts differing in their moral properties. This applies to cases in which: (a) the interacting agents decide their course of action on the basis of a correct deliberation of their strong moral reasons for action and are disposed to act in accordance with the moral requirements of both situations; (b) an agent's behaviour reflects the properties of situations such that a difference between the actions in two situations entails a difference between the properties of the situations; and (c) the situations compared are identical in all properties except in the set of properties distinguishing acting inside cyberspace from acting outside cyberspace – i.e., in the set of properties distinguishing computer-mediated interaction from face-to-face interaction.

8.2.1   Foundation for explaining [TBP]

Throughout the book a number of steps have been taken in order to be able to explain the basic premise. Of direct relevance for the explanation to be provided here, it has been shown that:

1. An agent's belief in the determinateness of the patient plays an important role for the agent's formation of the belief in the reality of the particular patient (6.1).

2. An agent's belief in causally affecting the patient is sufficient for the formation of an agent's belief in the reality of the particular patient, and an agent's belief in the effect of her actions on the patient being constitutive of her life-world adds to the strength of the agent's belief in the reality of the particular patient (6.2).

3. An agent's belief in the vulnerability of the patient is a necessary prerequisite of the belief in the reality of the particular patient (6.3).

4. Human agents generally strive to justify their beliefs in the sense of supporting them by means of evidence (6.4), and furthermore human agents generally seek to justify beliefs that may be justified by observational evidence by means of observational evidence (7.1).

5. On the basis of observational evidence an agent may arrive at the belief in the determinateness of the patient, at the belief in having a causal effect on the patient and at the belief in this causal effect on the patient being constitutive of the agent's life-world and of the vulnerability of the patient (7.1).

6. Computer mediation of interaction implies that there are limitations on the data and information exchanged in such interaction (4.2), and by extension that less evidence is available than in face-to-face interaction (7.1).

7. Interaction in chat-rooms and in certain kinds of tele-operation implies to varying degrees the loss of sensory access to the patient and to the setting (4.2), and hence the loss of evidence indicative of the properties of the patient and setting (7.1).

8. The loss of evidence indicative of the properties of the patient and the setting is a loss of evidence relevant for the formation and justification of the belief in the determinateness of the patient, of the belief in having a causal effect on the patient, of the belief in this causal effect on the patient being constitutive of the agent's life-world and of the belief in the vulnerability of the patient (7.1).

9. Interaction in chat-rooms and in certain kinds of tele-operation offers anonymity in two important senses of the agent being able to conceal identity (4.2).

10. Human agents occasionally experiment with their identity, and the anonymity offered in chat-rooms (and tele-operation) allows for extensive experimentation with identity (7.1).

11. Human agents are to a certain extent equipped with a Hobbesian psychology, which means that they may under conditions of anonymity choose to act so as to maximize their utility by exploiting others (7.1).

12. Human agents are generally aware of the possibility of others experimenting with their identity and of others acting so as to exploit them under conditions of anonymity, and they therefore have reason to rely less on the information exchanged, and thus on the evidence exchanged regarding the properties of the patient and the setting (7.1).

13. An agent believing she is required – or simply is motivated - to extend moral concern to a patient will change this belief – or lose her motivation – if she comes to believe that the particular patient towards whom the moral concern is directed is not real: i.e., that the patient does not possess the set of properties ascribed to her (8.1).

14. The moral practice of human agents is in general a response to the particularities of situations and of the people in it: i.e., it does not take others into account merely as human beings (8.2).

8.2.2   Explaining [TBP]

Having listed all the results of the preceding chapters that have direct relevance for the explanation of [TBP], let us now move on to provide the complete explanation. Briefly, we have to explain that interaction in cyberspace occasionally differs from face-to-face interaction by virtue of being constituted by agents performing acts differing in their moral properties. Applying the model [Exp], the explanation of [TBP] starts by outlining how an agent may fail to extend moral concern to an interacting party in cyberspace:

(**Exp**$_{Cyb}$)   (a)   Agent A holding the set of beliefs B was interacting with a patient in a chat-room (tele-operation).

(b)   Interaction in a chat-room (tele-operation) is characterized by limitations on the exchange of data and information, reduced sensory access to others and anonymity.

(c)   Because of the properties listed in (b) agent A lacked relevant observational evidence indicative of the properties of the patient and of the setting.

(d)   Agent A was disposed to justify certain beliefs on the basis of observational evidence.

(e)   Because of the properties listed in (b) and certain beliefs $b_1, b_2 \ldots b_n \subseteq B$ agent A came to believe that the available evidence indicative of the properties of the patient and of the setting was not reliable.

(f)   Because of (c), (d), (e) and certain beliefs $b_1, b_2 \ldots b_n \subseteq B$ agent A did not come to hold beliefs in the determinateness of the patient, in having a causal effect on the patient, in this causal effect being constitutive of A's life-world, and in the vulnerability of the patient.

(g)   Because of (f) agent A came to doubt the reality of the particular patient.

(h)   Agent A was motivated to extend moral concern to the patient by doing p in the relevant situation only if the particular patient was believed to be real

(i)   Because of (g) and (h) the agent A was not motivated to extend moral concern to the patient by doing p.

This explanation of why an agent A did not extend moral concern in interacting with a patient in a chat-room requires some comment. First, it must be noted that the beliefs, $b_1, b_2 \ldots b_n \subseteq B$, referred to in premise (e), are beliefs both in the dual possibility that the patient is experimenting with identity and acting so as to maximize her utility in and through the exploitation of others, and also those beliefs allowing the agent to infer from these possible actions of the patient that the evidence gained is unreliable. Second, and related to this, it must also be noted that the beliefs, $b_1, b_2 \ldots b_n \subseteq B$, referred to in premise (f), are the beliefs determining what is of relevance for the formation and justification of the beliefs in the determinateness of the patient, the belief in having a causal effect upon the patient, the belief in this causal effect being constitutive of the agent's life-world and the belief in the vulnerability of the patient.

Third, it must be noted that we have included in the explanation premises that are not, strictly speaking, necessary for the explanation of why the agent did not extend moral concern by doing p in interacting with a patient in a chat-room. These premises are included in order to reveal the links between the premises of the explanation as they have been disclosed throughout the book, and hence to increase our understanding of why the agent acted in the relevant way. As stated at the outset, the focus on explanation in this book has served exactly the purpose of showing in detail how the change in the setting from outside to inside cyberspace may come to influence behaviour.

Fourth and finally, it must be noted that the explanation provided above does not in itself imply that the agent A failed to fulfil a moral requirement. Thus the fact that agent A came to doubt the reality of the particular patient – i.e., the properties ascribed to the interacting agent – and consequently lacked the motivation to extend moral concern to the patient, may happen to coincide with the unreality of the particular patient – i.e., the fact that the patient did not actually possess the set of properties ascribed to her. In that case the lack of motivation to extend concern to the patient on the basis of the appearance of the patient is not a failure in the sense of failing to satisfy what the agent would consider a moral requirement. However, this would mean that the fact that the agent does not fail to satisfy such a moral requirement is merely coincidental. Moreover, in the light of the overall aim of this book to provide an explanation of the assumed difference between the ethical character of interaction inside and outside cyberspace, this coincidence may be ignored. Thus it is still the case that agents would occasionally fail to extend moral concern to patients in situations where they would have believed themselves to be under a moral requirement – for the reasons given in the explanation above. The explanation provided above is therefore still relevant in the attempt to explain the basic premise. Let us consequently move on to complete the explanation of the basic premise ([TBP]).

The remaining part of the explanation of the difference between interaction inside and outside cyberspace given in [TBP] is fairly straightforward. Throughout our investigations of the key properties of interaction in cyberspace we have more or less explicitly contrasted these with the properties of interaction outside cyberspace: that is to say, the properties of interaction face-to-face. In short, the conclusion is that in face-to-face interaction an agent has a higher degree of relevant and reliable observational evidence indicative of the properties of the patient. Hence an agent has more observational evidence on the basis of which she may form and justify her beliefs in the determinateness of the patient, the belief in having a causal effect on the patient, the belief in this causal effect being constitutive of the agent's life-world and the belief in the vulnerability of the patient – and, in turn, the agent may thus be claimed to have the evidence required in order to form a strong belief in the reality of the particular patient. Hence the agent will, other things being equal, be motivated to extend moral concern to the patient where, according to the agent's own moral beliefs, this is required. The difference between actions performed by an agent in interaction inside and outside cyberspace comes to be explained, therefore, by the fact that in certain kinds of computer-mediated interaction agents may arrive at a belief in the unreality of the particular patient whereas the same agents will apparently not form this belief in face-to-face interaction.

To be a little more specific, the basic premise ([TBP]) states that the difference between interaction inside and outside cyberspace is a difference between the moral properties of actions that seems to appear when two conditions are fulfilled. The first is when the interacting agents decide their course of action on the basis of a correct assessment of their strong moral reasons for action and are disposed to act in accordance with the moral requirements of both situations. The second is when an agent's behaviour reflects the properties of situations in such a way that a difference between the actions in two situations entails a difference between the properties of the situations. Has this more specific relationship been explained? The answer is affirmative.

Concerning the assumed difference between the moral properties of the actions performed, it is clearly the case that there will be such a difference between the actions performed if an agent in interaction in cyberspace does not satisfy a moral requirement that they would satisfy outside cyberspace. Suppose, for instance, that an agent holds certain ethical principles implying that the discouraged ought to be verbally encouraged. Compare now the situation of the agent having a conversation face-to-face with that of the agent having a conversation in a chat-room, and imagine that the patient in both situations says, 'I am so utterly discouraged'. It seems plausible to suppose that the application of the ethical principle adhered to by the agent would in both situations require the same action: for example, the exhortation to 'cheer up'. If the agent fails to satisfy this moral requirement in one situation and succeeds in the other, there is clearly a difference between the moral properties of these actions, or in other words a difference between those properties that determine whether or not the action satisfies the relevant ethical principle.

[TBP] specifies that the difference in interaction is supposed to show in cases where agents decide their course of action on the basis of a deliberation of their normative reasons for action, and the explanation provided takes this specification into account. Thus the agent A, figuring in the explanation above, is an agent who is striving to form and justify beliefs on the basis of relevant and reliable evidence and who is motivated to extend moral concern on the basis of a deliberation of the properties of the particular patient and situation as a whole.

Finally [TBP] specifies that the difference in interaction is supposed to reflect a difference between the two situations, which would have to be among the limited set of properties distinguishing interaction inside from interaction outside cyberspace. This requirement is clearly met since the explanation provided trades on a difference distinguishing the situations of interaction inside and outside cyberspace, namely the difference in the sensory access to the patient and hence in the observational evidence indicative of the state of the patient.

Having thus ensured that the explanation provided here fits the specifics of the premise ([TBP]) with which we embarked on this journey, we are now in a position to conclude our investigations. The task of providing an explanation of the basic premise ([TBP]) has been completed.

8.2.3 Revisiting sources of inspiration

Before we close this chapter, it may be worth putting the result into perspective. From the very first stages of this book we have been preoccupied with the attempt to provide an account of the role played by the bodily face of another person in the ethical character of interaction that was an alternative to that of Levinas, a detailed and systematic account that would explain a difference between the ethical character of interaction inside and outside cyberspace which the 'Legal Tender' experiment gave reason to believe in. Being at the end of this journey, we are now able to paint a picture of interaction that captures the Levinasian insight into the ethical importance of the face of another person in a way that makes it explanatory of the alleged difference in the ethical character of interaction.

The picture looks like this: in interaction we build beliefs concerning the attributes or properties of the person with whom we are interacting, and we hold these beliefs with varying degrees of certainty; in other words, we are convinced to varying degrees that other people are in actual fact constituted by the set of properties ascribed to them – that they are real as particular agents. The claim in this book is that the formation of these beliefs and the certainty with which we hold them depend on the availability of relevant and reliable evidence – but not evidence alone. For one of the claims made in this book is that evidence is not simply evidence. Evidence of certain matters plays a crucial role in the formation and strength of our belief in the reality of an agent as a particular agent. The role of the face is linked to the role of this evidence. Thus the face of another person is an important source of the evidence by means of which we form and trust in our beliefs concerning the reality of another person. Losing perceptual access to the face of another person – losing perceptual access to the physical presence of another person – is losing a crucial source of evidence of certain matters influencing the formation and strength of our belief in the reality of another person. In certain kinds of interaction in cyberspace we not only lose relevant evidence, but the evidence available is also for several reasons considered less reliable. The conditions for building and trusting our beliefs concerning the reality of the particular other person are, therefore, far from favourable. The ethical relevance of the change in epistemological conditions when moving from face-to-face interaction to interaction in cyberspace has to do with a basic precondition of extending moral concern to other people. Thus it seems as though we generally extend moral concern to others only if we do not doubt that they are what they seem to be: i.e., that they are real as particular agents. This in turn has to do with another feature of moral agency: namely, that it generally seems to be sensitive to the particularities of other people. Ultimately, then, the loss of access to the face – the body and setting – of another person means that a precondition of extending moral concern to another person is not met, and hence moral concern is not extended.

This picture of the ethical role of the face of another person is the main achievement of this book. As we will suggest in the next chapter, there are other plausible explanations of the alleged difference between the ethical character of interaction inside and outside cyberspace. From the very outset we have emphasized, however, that the role of explanation has been as a heuristic device enabling a detailed study of the ethical role of the face of another person, a study that would infuse the face of another person with an importance sufficient to influence behaviour. This has now been done.

Chapter 9

Concluding remarks

The investigations in this book into the mechanics of interaction in cyberspace may have raised some objections and perhaps alternative suggestions as to how the apparent difference between the ethical character of interaction inside and outside cyberspace is to be accounted for. This, the final chapter of the book, points to some of the alternative ways of approaching the basic premise.

The chapter serves the purpose of broadening the perspective. This book clearly has as its focus the attempt to understand how a belief in the reality of a particular person is constituted and may influence action. However, as already stated, throughout the writing of a book such as this many ideas and thoughts of differing degrees of relevance come to mind.

T. Ploug, *Ethics in Cyberspace: How Cyberspace May Influence Interpersonal Interaction,*   **205**
© Springer Science+Business Media B.V. 2009

9.1   Alternative explanations and interpretations

In this book we have approached the basic premise ([TBP]) with the aim of providing an explanation that endows the face of another person with ethical importance. However, looking at the basic premise without such intentions, there seem to be quite a number of possible alternative explanations of it. In this section we will take a very brief look at a few of those roads we could have gone down in our attempt to explain the basic premise. Each of these suggested alternative explanations is accompanied by a short comment as to how it fits into the overall framework of this book.

Inadequate understanding

One alternative explanation may draw on the lack of sensory access to the interacting party already shown to hold for certain kinds of interaction in cyberspace. Thus it could be argued that the lack of perceptual access may come to cause a lack of understanding or awareness of the moral requirements of the situation in the relevant kind of interaction in cyberspace, which, in turn, causes agents occasionally to act in ways that do not satisfy the moral requirement. Although this explanation clearly draws on the same features of interaction in cyberspace as the explanation already given – the lack of sensory access – such an explanation differs from the explanation given. The lack of sensory access causes a lack of understanding or awareness of the moral requirements, whereas the lack of sensory access in the explanation provided throughout this book is the cause of a lack of a belief in the determinateness or vulnerability of the patient, in having a causal effect on the patient and in this effect being constitutive of the agent's life-world, which is then taken to be the cause of a lack of a belief in the reality of the particular patient.

It may be established straight away that this explanation seems to have a limited applicability. Thus it seems as though in many cases an interacting party, in and through the words constituting a chat, may provide the information required for the agent to determine the moral demands of the situation and hence to understand and be aware of the moral demands of the situation. On the assumption that there may still be a difference in behaviour as specified in [TBP], the explanation is clearly inadequate.

It may be added that, as soon as we complicate this picture by taking into account the lack of reliable evidence in cyberspatial interaction, the simplicity of this alternative explanation becomes less attractive. Understanding and believing in a moral requirement are then a matter not only of getting evidence of the particularities of the interaction but also of establishing the reliability of this evidence. In this book we have claimed that evidence of certain matters plays a crucial role in the strength of our beliefs concerning the particularities of a patient. In so doing, we have provided an answer to the question facing this alternative explanation of the assumed difference in interaction.

Unfamiliarity and alienation

A second alternative explanation of the assumed difference between interaction inside and outside cyberspace may draw on the apparent widespread lack of familiarity with technological means of interaction. In this approach *some* people would be taken to act morally differently because they are unfamiliar with interaction using technological means. A related but slightly different explanation would see the difference in interaction as the outcome of the technological mediation of interaction simply alienating the interacting parties from each other. In this approach *all* people would potentially be acting differently inside cyberspace from outside cyberspace owing to technological equipment creating some sort of alienation between the interacting parties. Both of these explanations may actually refer to beliefs of interacting parties concerning the reality of the particular other. Thus the alienation may be taken to be somehow the cause of a belief in the unreality of the particular interacting party and so on.

Our response to this suggestion is twofold. We will readily admit that all kinds of technology seem to have some sort of 'alienating effect' – especially at the time when they are introduced. Without exploring these considerations further, it has to be noted that we may account for this feature of technology-mediated interaction within the framework provided here. Thus, insofar as the alienation is, so to speak, something happening in the deliberative mind, it may be interpreted as the fact of an agent coming to doubt the reality of the particular interacting party as a consequence of lacking sensory access to the patient. However, if the alleged alienation is something happening as a consequence of deeper psychological traits, then we are clearly unable to account for it – but then it also clearly falls outside the scope of this philosophical investigation into the consequences of moving from interaction outside cyberspace to inside cyberspace.

A Hobbesian psychology revisited

Third, we must consider why the explanation of the difference in interaction does not run along Hobbesian lines. That is, if human beings are equipped with a Hobbesian psychology as outlined previously, and if interaction in cyberspace is conducted under a high level of anonymity, then the difference in interaction may be seen as simply reflecting our Hobbesian psychology of maximizing our utility at the expense of others in conditions of anonymity.

Initially this suggestion does not seem too far-fetched. As already argued, it seems as though we are, at least to a certain extent, equipped with a Hobbesian psychology, and in conditions of anonymity this may lead us to act in such a way as to exploit, and thereby harm, other people with whom we are interacting. However, it seems as though more than this is involved in some of the cases discussed throughout this book. In the introduction we referred to a number of newspaper articles that claimed – in a tacit comparison with ordinary face-to-face conversation – that the language used in chat-rooms is very crude and offensive. Assuming this to be true, we do not seem to be able straightforwardly to take for granted that the reason for this change in the character of the communication lies in the combination of, on the one hand, agents entertaining a preference for the abuse of the interacting party – or some other preference which may be satisfied by abusing another person – and, on the other, the anonymity afforded by cyberspace. It seems to be just as plausible that this difference reflects the fact that the agent doubts the reality of the particular interacting party.

Even if it may be proved that we occasionally entertain preferences for the abuse of other people or preferences that in turn involve the abuse of other people, then an explanation of the assumed difference in interaction would clearly apply only to cases in which we actually entertained such a preference. It is not clear that we would always entertain such a preference. The explanation suggested in this book could be taken to account for these 'exceptional' cases.

A less defensive response to the challenge from the Hobbesian approach to the assumed difference in interaction may take both the Hobbesian approach and the account provided in this book to be part of the explanatory story. We may to a certain extent entertain preferences conflicting with the preferences of others, and we may to a certain extent seek to maximize the satisfaction of our preferences – but this is not the full story of why we come to act differently inside and outside cyberspace. Part of the story has to do with the lack of evidence supporting those beliefs underlying our belief in the reality of the particular interacting party. When an agent acts differently inside and outside cyberspace, it is as a result of a complex conglomerate of factors, among which can be counted both our Hobbesian psychology and our beliefs concerning the reality of a particular other person.

A more 'offensive', 'Levinasian' response to the challenge from the Hobbesian approach may claim that this approach becomes relevant only in those situations in which the face of another person is absent. More specifically, one could argue that a preference for, say, the abuse of another person or some other preference which may be satisfied by abusing another person originates in the absence of a belief in the reality of the particular interacting party: in other words, in the absence of a belief in the ability to affect the patient causally, in the absence of a belief in the effect on the patient being constitutive of the agent's life-world and, finally, in the absence of a belief in the patient being vulnerable. In short, the dominance of our Hobbesian psychology in interaction is initiated by the lack of our conviction in the reality of a particular other person. This response to the challenge from the Hobbesian approach entails the mechanisms of interaction claimed in this book being primary in terms of explaining the difference in interaction. A Hobbesian approach merely adds another element to the explanatory chain.

Whether or not the suggested responses to the challenge from a Hobbesian approach are plausible seems to depend on further investigation. No doubt the Hobbesian approach has much to offer for our understanding of interaction. In this book, however, other trains of thoughts and ideas have been pursued, and we have tried to show how these notions may also be relevant in accounting for the difference between interaction inside and outside cyberspace. As should be clear, the driving force behind this choice of focus has to do with the influence of Levinas. In any case, while a Hobbesian approach to the explanation of the basic premise ([TBP]) seems to be worth further investigation, it falls outside the scope of this book.

Proximity and presence

Fourth and finally, we also have to touch on the question of why the concept of 'proximity' or 'presence' does not figure in our explanation of the basic premise. It may seem as though what is really at stake in interaction in cyberspace as compared to interaction face-to-face is the 'presence' or 'proximity' of another person. Hence when people act differently in cyberspace from in face-to-face interaction, then it is because of the difference between the 'presence' or 'proximity' of the interacting party in the two situations.

This reference to the 'presence' and 'proximity' of another person may even be combined with the hypotheses presented in this book. Instead of the hypotheses claiming that our conviction concerning the reality of the particular patient is linked to our belief concerning the determinateness of the patient, in having causally affected the patient and in this effect being constitutive of our life-world and of the vulnerability of the patient, the hypotheses could be claiming that it is our sense of the 'presence' and 'proximity' of the patient that is linked to these further beliefs. We sense the 'presence' and 'proximity' insofar as we form these further beliefs. In the absence of relevant and reliable evidence in support of these further beliefs the patient is 'absent' or 'distant', and we come to act differently from how we would act when sensing the 'presence' and 'proximity' of the patient.

At first sight this reinterpretation of the hypotheses and the line of reasoning presented in this book seems fairly plausible. There are, however, at least two reasons for rejecting this reinterpretation. First, the link between moral agency and the 'presence' or 'proximity' of another person is far from obvious. It is not at all clear that the 'presence' and 'proximity' of the patient are required in order for us to extend moral concern to the patient. It is thus perfectly sensible to ask why 'presence' or 'proximity' should be a prerequisite for extending moral concern. We have, on the contrary, tried to show how a belief in the reality of a particular patient is a prerequisite for extending moral concern to the patient. A second and related point is that the reinterpretation of the hypotheses along the lines suggested seems to leave us with the conundrum of interpreting the notions of 'presence' and 'proximity'. If 'presence' and 'proximity' may be lost by mediating interaction by means of computers, then clearly they cannot be interpreted in terms of physical distance since agents interacting in cyberspace may be physically close to each other.

In the light of the problems of defining 'presence' and 'proximity' and of substantiating their relevance for moral action, the opposite reinterpretation may actually be suggested. That is, it may be that the hypotheses and line of reasoning presented in this book may be taken to account for the role of 'presence' and 'proximity' for the ethical character of interaction. Consequently 'presence' and 'proximity' are to be understood as the conditions making available relevant and reliable evidence in support of the beliefs underlying the belief in the reality of a particular patient – a belief that has been shown to be relevant for the ethical character of interaction. Whether or not this is a viable reinterpretation of the concepts of 'presence' and 'proximity' obviously cannot be settled on the basis of the rudimentary considerations here. Further investigation into the literature devoted to these topics is required. It is important to note, however, that the investigations in this book may be relevant to the understanding of these notions.

Each small step that increases our knowledge inevitably opens new vistas, new questions. Much research lies ahead, but we hope that this book may come to act as a source of inspiration for the investigative journey into the field of cyberspatial interaction.

Bibliography

1. Fred Adams. Knowledge. *in [27, pp. 228–236]*.

2. Gunther Anders. *Burning Conscience*. Rowohlt Verlag, Hamburg, 1961.

3. Aristotle. *The Nicomachean Ethics*. Cambridge University Press, Port Chester, NY, 2000.

4. Gert Balling (ed.). *Homo Sapiens 2.0*. Gads Forlag, Copenhagen, 2002.

5. Jeremy Bentham. *An Introduction to the Principles of Morals and Legislation*. Batoche Books, Kitchener, Ontario, 2000.

6. Frans A. J. Birrer. Applying ethical and moral concepts and theories to it contexts: Some key challenges and problems. *Proceedings of the Conference: Computer Ethics Philosophical Enquiry (CEPE 97)*, 1998.

7. Simon Blackburn. *Spreading the Word*. Clarendon Press, Oxford, 1984.

8. Albert Borgmann. Information, nearness and farness. *in [34, pp. 90–107]*.

9. Albert Borgmann. *Holding On to Reality*. The University of Chicago Press, Chicago, IL/London, 1999.

10. David Brink. *Moral Realism and the Foundation of Ethics*. Cambridge University Press, Cambridge, 1989.

11. J. Glenn Brookshear. *Computer Science: An Overview*. Pearson, Boston, MA, 2009.

12. Stuart Brown et al. (eds.). *Conceptions of Inquiry*. Routledge, London, 1981.

213

13. Terrell Ward Bynum. Computer ethics: Its birth and its future. *Ethics and Information Technology*, 3(2):109–112, 2001.

14. Robert J. Cavalier. *The Impact of the Internet on Our Moral Lives*. State University of New York Press, Albany, 2005.

15. Jonathan Dancy. *A Companion to Epistemology*. Blackwell Publishers, Oxford, 1992.

16. Jonathan Dancy. *Moral Reasons*. Oxford University Press, Oxford, 1993.

17. Jonathan Dancy. *Introduction to Contemporary Epistemology*. Oxford University Press, Oxford, 1995.

18. Donald Davidson. *Essays on Actions and Events*. Oxford University Press, Oxford, 1980.

19. Mauro Dorato. *Time and Reality*. Clueb, Bologna, 1995.

20. Fred Dretske. *Knowledge and the Flow of Information*. Cambridge University Press, Cambridge, 1981.

21. Hubert Dreyfus. *What Computers Still Can't Do*. MIT, Cambridge, MA, 1999.

22. Hubert Dreyfus. *On the Internet*. Routledge, London, 2001.

23. Stacey L. Edgar. *Morality and Machines: Perspectives on Computer Ethics*. Jones & Bartlett, London, 2002.

24. Jan Faye. *The Reality of the Future*. Odense University Press, Odense, 1989.

25. Luciano Floridi. Virtual reality. *in [27, pp. 40–61]*.

26. Luciano Floridi. Information ethics: On the philosophical foundation of computer ethics. *Ethics and Information Technology*, 1:37–56, 1999.

27. Luciano Floridi (ed). *The Blackwell Guide to the Philosophy of Computing and Information*. Blackwell, Oxford, 2004.

28. Philippa Foot (ed.). *Theories of Ethics*. Oxford University Press, Oxford, 1967.

29. Harry Frankfurt. Freedom of the will and the concept of a person. *in [100, pp. 81–95]*.

30. Raymond Gillespie Frey (ed.) and Christopher Heath Wellman. *A Companion to Applied Ethics*. Blackwell Publishers, Oxford, 2003.

31. David Gauthier. *Morals by Agreement*. Oxford University Press, Oxford, 1986.

32. Bernard Gert. Common morality and computing. *Ethics and Information Technology*, 1:57–64, 1999.

33. Edmund L. Gettier. Is justified true belief knowledge? *Analysis*, 23:121–123, 1963.

34. Ken Goldberg. *The Robot in the Garden*. MIT, Cambridge, MA, 2001.

35. Nelson Goodman. Seven structures on similarity. *in [36, pp. 437–446]*.

36. Nelson Goodman. *Problems and Projects*. Bobbs-Merrill, New York, 1972.

37. Donald Gotterbarn. The use and abuse of computer ethics. *Journal of System and Software*, 17(1):75–80, 1992.

38. Donald Gotterbarn and Simon Rogerson. Computer ethics: The evolution of the uniqueness revolution. *Proceedings of the Conference: Computer Ethics Philosophical Enquiry (CEPE 97)*, 1998.

39. Gordon Graham. *The Internet: A Philosophical Inquiry*. Routledge, London, 1999.

40. Tommy Grøn. Vi sviner hinanden på nettet. *Urban*, 1. sektion:1, 30th of November, 2004.

41. Jürgen Habermas. *Moral Consciousness and Communicative Action*. Polity Press, Cambridge, MA, 1992.

42. John Heil. Metafysik efter 1960. *in [62, pp. 303–348]*.

43. Michael Heim. *The Metaphysics of Virtual Reality*. Oxford University Press, New York, 1993.

44. Michael Heim. *Virtual Realism*. Oxford University Press, New York/Oxford, 1998.

45. Christopher Hitchcock. Probabilistic causation. *Stanford Encyclopedia of Philosophy*, http://plato.stanford.edu/archives/fall2002/entries/causation-probabilistic/, Fall 2002.

46. Thomas Hobbes. *Leviathan*. Thoemmes Continuum, Bristol, 2003.

47. Ted Honderich. *The Oxford Companion to Philosophy*. Oxford University Press, Oxford, 1995.

48. David Hume. *An Enquiry Concerning the Principles of Morals*. Liberal Arts Press, New York, 1957.

49. David Hume. *An Enquiry Concerning Human Understanding*. Clarendon Press, Oxford, 2000.

50. Deborah G. Johnson. Sorting out the uniqueness of computer-ethical issues. *Etica and Politica*, 1(2), 1999.

51. Deborah G. Johnson. *Computer Ethics*. Prentice-Hall, Upper Saddle River, NJ, 2000.

52. Immanuel Kant. *Grundlegung zur Metaphysik der Sitten*. Reclam, Stuttgart, 1994.

53. Immanuel Kant. *Kritik af den praktiske fornuft*. Hans Reitzels Forlag, Copenhagen, 2000.

54. John Ladd. Ethics and the computer world: A new challenge for philosophers. *Computers and Society*, 27(3):8–13, 1997.

55. Emmanuel Levinas. *Etik og uendelighed*. Hans Reitzels Forlag, Copenhagen, 1995.

56. Emmanuel Levinas. *Totalitet og uendelighed*. Hans Reitzels Forlag, Copenhagen, 1996.

57. Emmanuel Levinas. *Fænomenologi og etik*. Gyldendal, Copenhagen, 2002.

58. Pierre Levy. *Collective Intelligence*. Perseus Books, Cambridge, MA, 1997.

59. Pierre Levy. *Becoming Virtual*. Plenum Trade, New York/London, 1998.

60. Pierre Levy. *Cyberculture.* University of Minnesota Press, Minneapolis, MN, 2001.

61. Kasper Lippert-Rasmussen. *Deontology, Responsibility, and Equality.* Museum Tusculanum Press, Copenhagen, 2005.

62. Poul Lubcke (ed.). *Engelsk og amerikansk filosofi — Videnskab og sprog.* Politikens Forlag, Denmark, 2003.

63. Alasdair MacIntyre. *Dependent Rational Animals.* Open Court, Chicago/ La Salle, IL, 1999.

64. Alasdair MacIntyre. *After Virtue.* Duckworth, London, 2002.

65. John Mackie. Causes and conditions. *in [89, pp. 33–55].*

66. John Mackie. *The Cement of the Universe.* Clarendon Press, Oxford, 1974.

67. Walter Maner. Is computer ethics unique? *Etica and Politica,* 1(2): 1999.

68. Charles Burton Martin. Substance substantiated. *Australian Journal of Philosophy,* 58:3–10, 1980.

69. Richard O. Mason. Applying ethics to information technology issues. *Communications of the ACM,* 38(12):55–57, 1995.

70. David McNaughton. *Moral Vision.* Blackwell Publishers, Oxford, 1988.

71. James H. Moor. What is computer ethics. *Metaphilosophy,* 16(4):266–275, 1985.

72. James H. Moor. The future of computer ethics: You ain't seen nothin' yet. *Ethics and Information Technology,* 3(2):89–91, 2001.

73. Peter Øhrstrøm and Per Hasle. *Temporal Logic.* Kluwer, Dordrecht, 1995.

74. Peter Øhrstrøm (ed). *IT-etiske temaer.* Syddansk Universitet, Kolding, 2003.

75. Derek Parfit. *Reasons and Persons.* Oxford University Press, Oxford, 1984.

76. Plato. *The Republic.* Ebrary, Kessinger, 1992.

77. Mark Platts. *Ways of Meaning.* Routledge Kegan Paul, London, 1979.

78. Mark Platts. *Reference, Truth and Reality.* Routledge Kegan Paul, London, 1980.

79. Thomas Ploug. Rationalitet og retfærdighed. *Filosofiske Studier*, 21:184–224, 2001.

80. Thomas Ploug. Relationer mellem semantik og ontologi i en temporal kontekst. *Filosofiske Studier*, 22:118–151, 2002.

81. James Rachels. *Elements of Moral Philosophy*. McGraw-Hill, New York, 1993.

82. John Rawls. *A Theory of Justice*. Oxford University Press, Oxford, 2000.

83. Howard Rheingold. *The Virtual Community*. MIT, Cambridge, MA, 2000.

84. Roger Scruton. *Kant: A Very Short Introduction*. Oxford University Press, Oxford, 2001.

85. Peter Singer (ed.). *A Companion to Ethics*. Blackwell, Oxford, 1993.

86. James Slevin. *The Internet and Society*. Polity Press, Cambridge, MA, 2000.

87. Michael Smith. *The Moral Problem*. Oxford University Press, Oxford, 1994.

88. Lawrence Snyder. *Fluency with Information Technology: Skills, Concepts & Capabilities*. Pearson, Boston, MA, 2008.

89. Ernest Sosa and Michael Tooley (eds.). *Causation*. Oxford University Press, Oxford, 1993.

90. Richard Spinello. *Cyberethics: Morality and Law in Cyberspace*. Jones & Bartlett, London, 2000.

91. Derek Stanovsky. Virtual reality. in *[27, pp. 167–177]*.

92. Eric Steinhart. Emergent values for automatons: Ethical problems of life in the generalized internet. *Ethics and Information Technology*, 1:155–160, 1999.

93. Jonathan Steuer. Defining virtual reality: Dimensions determining telepresence. *Journal of Communication*, 4:73–93, 1992.

94. Michael Stocker. Desiring the bad: An essay in moral psychology. *Journal of Philosophy*, 76:738–753, 1979.

95. Emil Møller Svendsen. Elever sviner lærere til på nettet. *Urban*, 2. sektion:1, 2nd of March, 2006.

96. Alan Turing. Computing machinery and intelligence. *Mind*, 59:433–460, 1950.

97. Sherry Turkle. *Life on the Screen: Identity in the Age of the Internet*. Touchstone, New York, 1995.

98. M. J. Van den Hoven. Computer ethics and moral methodology. *Metaphilosophy*, 28(3):266–275, 1997.

99. Gary Watson. Free agency. *in [100, pp. 96–110]*.

100. Gary Watson (ed.). *Free Will*. Oxford University Press, Oxford, 1982.

101. Bernard Williams. Internal and external reasons. *in [102, pp. 101–113]*.

102. Bernard Williams. *Moral Luck*. Cambridge University Press, Cambridge, 1981.

103. Bernard Williams. *Ethics and the Limits of Philosophy*. Harvard University Press, Cambridge, MA, 1985.

104. Ludwig Wittgenstein. *Philosophical Investigations*. Basil Blackwell, Oxford, 1974.

105. James Woodward. Causation and manipulability. *Stanford Encyclopedia of Philosophy*, http://plato.stanford.edu/archives/fall2001/entries/causation-mani/, Fall 2001.

106. Georg Henrik von Wright. On the logic and epistemology of the causal relation. *in [89, pp. 105–124]*.

107. Georg Henrik von Wright. *Explanation and Understanding*. Cornell University Press, Ithaca/New York/London, 1971.

Index

Printed by Publishers' Graphics LLC
DBT130910.15.14.96